The Humanized Internet

Dignity, Digital Identity and Democracy

RIVER PUBLISHERS SERIES IN COMPUTING AND INFORMATION SCIENCE AND TECHNOLOGY

The "River Publishers Series in Computing and Information Science and Technology" covers research which ushers the 21st Century into an Internet and multimedia era. Networking suggests transportation of such multimedia contents among nodes in communication and/or computer networks, to facilitate the ultimate Internet.

Theory, technologies, protocols and standards, applications/services, practice and implementation of wired/wireless networking are all within the scope of this series. Based on network and communication science, we further extend the scope for 21st Century life through the knowledge in machine learning, embedded systems, cognitive science, pattern recognition, quantum/biological/molecular computation and information processing, user behaviors and interface, and applications across healthcare and society.

Books published in the series include research monographs, edited volumes, handbooks and textbooks. The books provide professionals, researchers, educators, and advanced students in the field with an invaluable insight into the latest research and developments.

Topics included in the series are as follows:-

- Artificial intelligence
- Cognitive Science and Brian Science
- Communication/Computer Networking Technologies and Applications
- Computation and Information Processing
- Computer Architectures
- Computer networks
- Computer Science
- Embedded Systems
- Evolutionary computation
- Information Modelling
- Information Theory
- Machine Intelligence
- Neural computing and machine learning
- Parallel and Distributed Systems
- Programming Languages
- Reconfigurable Computing
- Research Informatics
- Soft computing techniques
- Software Development
- Software Engineering
- Software Maintenance

For a list of other books in this series, visit www.riverpublishers.com

The Humanized Internet

Dignity, Digital Identity and Democracy

Akram Alfawakheeri
The Humanized Internet, Germany

Monique J. Morrow
The Humanized Internet, Switzerland

River Publishers

Routledge
Taylor & Francis Group
NEW YORK AND LONDON

Published 2025 by River Publishers
River Publishers
Broagervej 10, 9260 Gistrup, Denmark
www.riverpublishers.com

Distributed exclusively by Routledge
605 Third Avenue, New York, NY 10017, USA
4 Park Square, Milton Park, Abingdon, Oxon OX14 4RN

The Humanized Internet: Dignity, Digital Identity and Democracy / Akram Alfawakheeri and Monique J. Morrow.

Routledge is an imprint of the Taylor & Francis Group, an informa business

ISBN 978-87-7022-032-3 (hardback)
ISBN 978-87-4380-073-6 (paperback)
ISBN 978-87-7004-799-9 (online)
ISBN 978-8-770-04798-2 (ebook master)

Dedication

To the millions of people worldwide who live unseen, whose stories go untold.

To those whose lives have been shaped by circumstances beyond their control, caught in the margins of an imperfect world. I hope this book offers a glimpse of hope that technology can open new doors to dignity and belonging.

To everyone who contributed to this book, and to all the quiet visionaries and tireless innovators using technology to uplift others – you are lighting the way toward a more compassionate future.

To my mother and my family, whose deep connection endures across oceans and continents, grounding me in love and strength.

And to Monique, my co-author, whose creativity and friendship have made this journey unforgettable. This book exists because of your belief and dedication.

Akram Alfawakheeri

For those who believe in the power of connection and the transformative potential of technology, this book is dedicated to the pioneers of the humanized internet, those individuals and organizations who envision a digital landscape that prioritizes humanity empathy, and inclusivity.

May we continue to strive for a future where technology uplifts and enriches the human experience, fostering a world that is as compassionate as it is innovative.

Thank you to those individuals who contributed to the book highlighting the inspiring use case examples that are so appropriate for the humanized internet.

To my co-author **Akram**, you became the story and center of why this book had to be written. Your story is a movie.

To River Publishers, I thank you all for your incredible patience and flexibility during this book project where I had been personally confronted with tragedy.

Thank you to my mentors and allies, who have dedicated their lives to making the internet a more human-centered place. Your influence inspires me every day.

I thank my friends and family for their support to me during this book journey. I am forever grateful for words of encouragement especially when I felt I could not continue.

Finally, this book is dedicated in loving memory of **Jacqueline V. Thévenaz** who passed away on August 11, 2023, succumbing to Stage IV Ovarian Cancer. "Veronique" embodied the values of the humanized internet with her intense curiosity as to "why" and "what-if." She was my source of continued inspiration and courage. She will always remain in my heart.

"Merci Véro!"

– **Monique Jeanne Morrow**

Contents

Foreword

In my book, *Heartificial Intelligence: Embracing our Humanity to Maximize Machines*[1], I point out how many people misinterpret the notion to, "Do No Evil" as espoused by Google's Founders in their actual 2004 IPO filing you can still read[2] on the internet today:

DON'T BE EVIL

> *Don't be evil. We believe strongly that in the long term, we will be better served—as shareholders and in all other ways—by a company that does good things for the world even if we forgo some short-term gains. This is an important aspect of our culture and is broadly shared within the company.*

There's no overt philosophical, moral, or ethical methodologies provided here to help define "good" or "evil" in any accountable fashion beyond the potential to "forgo some short-term gains." Along with the recognition that "we will be better served as shareholders" this phrase positions Google's vision within a global economic paradigm legally beholden to the individuals or corporations providing the financial basis for an organization to come into existence.

This means "good" in the context of this framing is defined by the prioritization of shareholder value as the Key Performance Indicators for a (what is widely known) organization that makes its primary revenue from advertising[3].

[1] *Heartificial Intelligence: Embracing our Humanity to Maximize Machines* by John C. Havens: Penguin Publishing Group, February 2, 2016. https://www.amazon.com/Heartificial-Intelligence-Embracing-Humanity-Maximize/dp/0399171711

[2] http://www.sec.gov/Archives/edgar/data/1288776/000119312504124025/ds1a.htm

[3] In the SEC filing for Alphabet Inc's (Google's) second quarter filings advertising revenues exceed Search and Cloud services, despite speaking points from leadership: https://www.sec.gov/Archives/edgar/data/1652044/000165204424000076/googexhibit991q22024.htm

As Alfawakheeri and Morrow note in their seminal and profound, *The Humanized Internet*, where humans don't have the ability to create an online identity or control the personal data generated based on their unique and individual existence, "search" or "the Cloud" or "GenAI" are not the products and services of our digitized and virtual modernity:

We are.

We're the product.

More specifically our singular human existence in digital, virtual, and spatial web realms is represented with data generated by technological services whose design and outputs we don't control and governments who legally mandate how we register our presence overall.

I first met Monique Morrow over ten years ago while she was a CTO for Cisco, a role demanding broad expertise in Internet and Data governance technologies in a global and interdisciplinary context. Hearing her speak at a conference focusing on Artificial Intelligence, I was immediately struck at the authority with which she could define complex technological and data governance issues with language and case studies providing immediate clarity and relevance for listeners.

As the Founding Executive Director for the IEEE Global Initiative on Ethics of Autonomous and Intelligent Systems, I approached her in 2016 to join our Executive Committee that created one of the earliest global compendiums focused on AI Ethics, *Ethically Aligned Design: A Vision for Prioritizing Human Well-being with Autonomous and Intelligent Systems*[4]. This document and work driven by Monique and hundreds of other global thought leaders also inspired the creation of the IEEE 7000™ standards series[5], the first standards of their kind in the history of IEEE to formally address societal and technical issues in context as a basis for the policy, design and use of technology.

While aggregating the various chapters for the 2019 version of *Ethically Aligned Design*, I approached my colleague Jay Iorio to note we really needed to address issues of Augmented and Virtual reality to fully address the ethics of AI and data governance. He agreed and recommended we approach Monique based on her deep expertise in those areas to co-chair a committee focused on these topics. This led not only to a chapter in *Ethically Aligned Design*, but the creation of *The IEEE Global Initiative on Ethics of Extended*

[4] The IEEE Global Initiative on Ethics of Autonomous and Intelligent Systems. *Ethically Aligned Design: A Vision for Prioritizing Human Well-being with Autonomous and Intelligent Systems*, First Edition. IEEE, 2019. https://sagroups.ieee.org/global-initiative/wp-content/uploads/sites/542/2023/01/ead1e.pdf

[5] https://standards.ieee.org/initiatives/autonomous-intelligence-systems/standards/#p7000

Reality[6] that has to date released eight white papers focused on privacy, governance, and business in what is now referred to as the Metaverse[7].

What I learned from working with Monique is what you'll begin to understand on myriad levels of consciousness as you read, *The Humanized Internet* – you and your data have worth you should be able to control. This is a core aspect of Monique's humanity that may be the only part of her personage that outshines her intellect – her kindness. As a thought leader in business, academic and policy arenas her authority and expertise are without question but it's in her empathetic and loving presentation of ideas she personifies the humanity this book provides, particularly in regards to understanding ideas of genuine consent and agency.

Having worked as a journalist and as an applied ethicist for more than twenty years I've come to recognize most people consider "genuine consent" as language framing byzantine length, privacy terms and conditions statement to be avoided via clicking an enforced form of acceptance if one wants to be a part of digital modernity.

That's just the legal aspect of genuine consent however, typically employed within a shareholder value paradigm designed to keep users from fully comprehending the vital importance of what they are agreeing to and how they're doing it.

Monique has been providing an understanding of this obfuscated and myopic dynamic for decades, reminding society at large that content regarding data ownership and accessibility is in fact *an issue of human rights*. Where any voice or agenda would counsel our data is worthless and can't be retrieved, she has provided a foundational voice in the world of Personal Data Governance and global policy to note that where information represents our identity, not valuing it is the equivalent of saying, "human trafficking is hard to address, so let's just let it keep happening."

Yet the trafficking of our data and identity is also immoral. Unethical. It is theft of the highest order where the digital vapor representing our unique selves is essentially our soul in silicon.

Monique and her illumination with Alfawakheeri in *The Humanized Internet* also provide a pragmatic basis for one of the most challenging concepts to elucidate in computational and virtual realms, which is agency. Often addressed directly in fields of Human Computer Interaction or Affective

---

[6] https://standards.ieee.org/industry-connections/activities/ethics-extended-reality/

[7] Terminology that all good geeks know was originated by Neal Stephenson in his groundbreaking book, Snow Crash versus Zuckerberg's updated name for Facebook: https://en.wikipedia.org/wiki/Metaverse.

Computing, agency essentially infers communication of any kind provided by context along or without the use of actual spoken language.

As an metaphor to understand this phenomenon, consider that nobody stands up and yells, "Hey, you're about to see a movie now so you're allowed to experience catharsis in a safe way based on the fictional experience of characters on the screen" when you go to the theatre because the lights being dimmed provides a form of agency about the nature of performance humans have become used to as a signal for safety. The boundaries of their experience are known, and they are safe to think, dream or react with tears or laughter in their seats.

The Humanized Internet provides a clear navigation of these types of signals not expressed by the governmental and technological powers driving Generative AI or data governance today. Building on my movie metaphor, Alfawakheeri and Morrow shine a light on how the possibility of data sovereignty is offered (or not) by governments and how companies definitively know the difference between manipulation and "nudging" and lean towards the former as a design choice to keep us clicking and consuming as the definition of success for the small cadre of shareholders allowing our existence as virtual vassals.

At least as long as we have money or credit to offer.

In 2011 Eli Pariser showed us in *The Filter Bubble: What the Internet is Hiding from You*[8], that our Internet experience (where "internet" here infers any form of interaction with a user or their data beyond physical-only existence) is predominantly shaped by our clicks framed more often towards purchase than purpose. In 2019 Shoshana Zuboff in *The Age of Surveillance Capitalism: The Fight for a Human Future at the New Frontier of Power*[9] further demonstrated the fiscal framing of our digital existence mired in centuries of physical colonial practices being modernized to gain control of our data, minds, and policy.

Continuing in this spirit of revolutionary revelation, Alfawakheeri and Morrow remind modern readers in the age of Generative AI that the foundation of all technologies demand transparent and accountable metrics of accountability versus vague moralistic intentionality:

> Moreover, "Do No Evil" should not be a company motto but rather foundational to the development of any technologies that do impact us

[8] https://www.amazon.com/Filter-Bubble-What-Internet-Hiding/dp/1594203008 (Please add correct citation here).

[9] https://www.amazon.com/Age-Surveillance-Capitalism-Future-Frontier/dp/1610395697 (Please add correct citation here).

as individual consumers of these technologies and corresponding products. Indeed, there is a polarity between an internet that is used for mass empowerment and one that can be used for mass destruction. Privacy, security, and management of your digital footprint should be done by YOU[10].

You are not a product. You are a person.

Unique, sacred and sovereign.

Your worth as represented by your data is yours to understand, learn from, share and own.

The Humanized Internet provides the background, permission, and roadmap for you to recognize these truths despite the propagandized business or government agendas telling you, "The horse has left the barn" and "our data is already out there."

Despite what may be posted, your life and serialized spirit are still yours to redeem and share.

For genuine consent and human agency to benefit current humans and future generations of children and the planet, we need *The Humanized Internet*, now.

John C. Havens
Author, *Heartificial Intelligence: Embracing our Humanity to Maximize Machines*

Founding Executive Director,
The IEEE Global Initiative on
Ethics of Autonomous and Intelligent Systems

[10] Please insert proper attribution for the book / page number where this appears here.

List of Contributors

Alfawakheeri, Akram, *The Humanized Internet, Germany*

Gasteiger, Daniel, *Co-Founder and Managing Director Saipient AG,Zürich, Switzerland*

Hughes, Frances, *The University of Sydney, Australia*

J. Morrow, Monique, *The Humanized Internet, Switzerland*

Mehran, Zarrebini, *PFE International Inc., United Kingdom*

Pradeep, Adithya Kumar, *CEO Founder, Jeev, Switzerland*

List of Figures and Table

List of Abbreviations

ACLU	Americans for Civil Liberties Union
AMIA	American Medical Informatics Association
CCPA	California Consumer Privacy Act
CHAMPION	CardioMEMS Heart Sensor Allows Monitoring of Pressure to Improve Outcomes in NYHA Class III Heart Failure Patients
CSR	Corporate social responsibility
DAO	Decentralized autonomous organizations
DBT	Direct benefit transfer
DID	Decentralized identifier
DIF	Decentralized Identity Foundation
EFF	Electronic Frontier Foundation
EPIC	Electronic Privacy Information Center
ESG	Environment social governments
FDA	Federal Drug Administration
GDPR	General Data Protection Regulation
GPS	Global positioning system
HIPAA	Health Insurance Portability and Accountability Act
ICD	Internal cardioverter-defibrillator
IETF	Internet Engineering Task Force
IIW	The Internet Identity Workshop
IP	Internet Protocol
IP	Intellectual property
ISP	Internet service provider
KYC-AML	Know your customer-anti money laundering
MVP	Minimum viable products
NEAT-HFpEF	Nitrate's effect on activity tolerance in heart failure with preserved ejection fraction
NGO	Non-governmental organization
NSTC	National Science and Technology Council
OWI	One world identity
PALM	Patient and provider assessment of lipid management

PII	Personally Identifiable Information
POD	Personal Online Data Store
PRO	Performing rights organizations
SMPC	Secure multiparty computation
SSI	Self-sovereign identity
SSL	Secure sockets layer
TDLID	Trusted Digital Lockbox ID
UGC	User generated content
W3C	World Wide Web Consortium

PART I

Setting the Framework for The Humanized Internet: Digital Identity, Human Dignity

DOI: 10.1201/9781003339724-1

Introduction

Vision of the Humanized Internet

When we review the history of the internet, no one envisaged that the internet would be used globally as it is now. The internet has become one of the most phenomenal inventions of the past 20th century. Whether it is email, instant chat messaging, web mechanisms and so on, the internet "just works." We are simply connected, thanks to the internet, and this communication is further enhanced via mobile capability. We have become tethered to our internet for instant gratification of responding and communicating in "real time." We can communicate in 240 characters and use social media platforms to showcase a birthday dinner, a marriage, or simply capture photographic moments in "selfie-mode." The internet has been an enabler of this need to share and narrate in real-time by many individuals. We prepare and receive text messages for shopping deliveries and yes, sit at dinner tables in restaurants or in trams, text messaging.

Indeed, the internet has become so ingrained in our daily lives from learning, commerce to the so-called "smart homes," and we are enveloped by the magic of this internet.

The polarity between that which can be used for "mass empowerment" and "mass destruction" is becoming more evident such that organizations are calling for a *Digital Geneva Convention.* Internet platform companies are refusing to work with government agencies in launching the so-called cyberattacks.

Yet, there is not a day that goes by that we read about data being hacked, inadvertently leaked or simply held for ransom. We simply are "not in control of our data."

While it is noble to observe the articulation of ethics by these companies, some cannot help but reflect on "benevolent abuses," after all we are all products, right?

DOI: 10.1201/9781003339724-2

As consumers of the internet, can we ask the question whether or not the internet is truly autonomous and not "controlled" by a few companies or organizations?

Within the Internet Research Task Force, there is a Human Rights Protocol Considerations "chartered to research whether standards and protocols can enable, strengthen, or threaten human rights, as defined in the *Universal Declaration of Human Rights (UDHR) and the International Covenant on Civil and Political Rights (ICCPR)*, specifically, but not limited to, the right to freedom of expression and the right to freedom of assembly."

The concern for the so-called "algorithmic decision-making and human rights" is further evidenced by the corpus of research and the emergence of work such as Ethically Aligned Design from the IEEE Global Initiative on Ethics of Autonomous and Intelligent Systems.

Quo vadis our Internet?

While reading this book, there are key themes that are constant such as the notion of identity and identity sets, e-sovereignty and privacy, and, most importantly, the function of an internet that is inclusive, not "controlled" by a few organizations for their own profitability. Certainly, "enterprising" the internet has been a process over these past years, and there is no intent to set judgment here but rather pause for a moment and reflect on the impact of these technologies to individuals. Yes, this is the Humanized Internet. These tenets may sound libertarian but in fact we are speaking about core principles to guide the development and perhaps the return of the internet to the people, especially those who are underserved.

At a macro level, UN SDG Goal 16.9 [1] has identified the goal to provide a legal identity to all including birth registration by 2030.

While this goal is visionary, the implications of attaining such a goal beg several questions with trust being foundational. If we bind the problem area to portability of digital identities and to trusted attestation of digital credentialing, what can be done in developing a trust framework for citizens and for policy-makers?

The assumption *ceteris paribus* is that the citizen be the center of this trust universe. A further motivation for writing this book with like-minded leaders has been around a societal trust deficit overall.

There has been prior work in this space. [2]

There are various questions to consider when correlating trust deficit with digital identity.

1. Do we have a trust deficit when speaking about the mirroring of our physical identities with our digital identities?

2. How do we define trust as per point 1 above?
3. What is the ecosystem of players in the trust universe as we emerge more and more in the digital space?
4. What is the polarity between trust, privacy, and cybersecurity when referring to digital identities and the so-called portability of these identity sets?
5. What standards exist? What standards must be developed?

There is a body of research that seeks to define "trust" which is a complex sociological, psychological concept and the basis for human interactions. [3] As the old adage goes, "Trust takes years to build, seconds to break, and forever to repair." This fact is described in the Netflix documentary entitled, "The Great Hack," where the privacy and abuse of individual/citizen data is the tip of an iceberg of a calculated strategy to take both data and control away from the individual. [4] An observation has been that the "outrage" from these data abuse has not come from individual/digital citizens but rather from nation-state governments and privacy, human rights organizations. [5] This fact begs the question as to whether or not the individual is immune to such data abuses that involve his/her data. Perhaps the response to this hypothetical question may be the degree of perceived or actual abuse, for example, personal bank account and funds stolen versus identity fraud versus data leakage with no "direct harm" as a result of an organizational breach.

If the degree of perceived or actual abuse is greater than "no direct harm," then the question is whether or not the egregious act has resulted in the individual behavioral change from passive to active digital citizen. There is an abundance of research on human hacking, but in terms of empirical and behavioral analysis that confirms a change as a result of such abuse, the research is lacking at least at a macro level. [6] This particular question may be difficult to ascertain because individuals (collectively) may be disinclined not to share the fact that they have been victims. Therefore, the postulation is that behavioral change is possible when correlated with the severity of the abuse.

However, trust can manifest itself in a Thomas Hobbes like Leviathan where a central authority like a government entity operates in the background to assure that society behaves according to the rules that have been developed by the central authority. [7] As social psychologist Thomas Tyler observed, rules will be followed if the citizens perceive procedural fairness. [8] At the opposite end of this trust spectrum is peer-to-peer (P2P) trust, which is based largely on relationships and shared ethical norms. P2P trust may be interpersonal as experienced among families, friends, or via a shared set of operating principles, for example, the internet. Furthermore, the modality of the engineers who develop protocols and architectures to evolve the internet, i.e., the Internet Engineering Task Force or (IETF) is "rough consensus and running code" and "we reject kings, presidents, and voting." [9]

Moreover, "Do No Evil" should not be a company motto but rather foundational to the development of any technologies that do impact us as individual consumers of these technologies and corresponding products. Indeed, there is a polarity between an internet that is used for mass empowerment and one that can be used for mass destruction. Privacy, security, and management of your digital footprint should be done by YOU.

With the progression of human and machine interaction due to advances in Biotech, brain–computer interface, cloud computing, virtual reality, and mixed reality, we need to understand the impact of these technologies to identity overall. Do we require a new definition of identity? What is e-sovereignty or distributed sovereign identity and its application moving forward if we posit that the institutions that exist today may indeed no longer be relevant in their current structure. We have read about the abuses when our data fall into the hands of other entities, intentionally or not.

The Humanized Internet is therefore a call to action, your action as the reader.

The vision of the Humanized Internet may be broken down into three core messages:

1. **The first message is around the liberty of identity and a digital instantiation therein.** Imagine you live in a country where there is war, and you have to leave immediately. Or imagine all your valuable documents have been lost in an earthquake or fire. Could we envision a distributed digital lockbox where we as individuals possess the digital keys?

2. **The second message is on control: Are you in control over your identity, and do you care?** Every day a lot of identities are hacked and/or misused by the very organizations that citizens trust. There is a general malaise that has set in with individuals asserting either 'that won't happen to me' or 'it is already too late'. We do believe that it is not too late, and yes, your digital identity can be abused.

3. **The third message is within ethics and technology:** As individuals, we are all being profiled through technology. Take smart homes for example: These electronic products are listening, watching you. But who is listening? The potential for misuse is very strong. How can we

define the intentional use of technology recognizing and responding to cognitive bias therein?

This observation begs the question as to how a chasm may be resolved and by which entities. What kind of social contract is required for the digital world we share? In the 21st century, fostered by the use of social media and the internet, social contracts have been evolving to identity politics and the notion of multiple identities, especially when referring to IoT and to the Internet of Everything and Everyone. Dignity and the struggle for recognition are provoking the requirement for a new social contract that can function in a digital world. Given the pattern of decentralization or a hybrid therein, a novel and relevant social contract must put the citizen in the center of the digital universe as an active player.

As you, the reader, embark on the journey of the Humanized Internet, we would like to chart out the main parts of the book to guide you.

This book is divided into four main sections:

Part I: Setting the Framework for The Humanized Internet: Digital Identity and Human Dignity

Part II: Societal Dilemmas

Part III: The Humanized Internet Enablers

Part IV: Evoking the Humanized Internet Use Cases

Conclusion

Every section and chapter is designed to provoke action in the form of recommendations. We hope you enjoy this journey and moreover become active in the Humanized Internet.

References

[1] https://sustainabledevelopment.un.org/?menu=1300
[2] https://www.openidentityexchange.org/wp-content/uploads/2017/06/OIX-White-Paper_Trust-Frameworks-for-Identity-Systems_Final.pdf
[3] Robbins, B.G. (2016). What is Trust? A Multidisciplinary Review, Critique, and Synthesis. *Sociology Compass* 10(10), 972–986. 10.1111/soc4.12391.
[4] *The Great Hack*. (2019). Available at: https://www.netflix.com/Title/80117542 (accessed 26 July 2019).

[5] Isaak, J., and Hanna, M.J. (2018). User Data Privacy: Meta, Cambridge Analytics and the Privacy Protection. *Computer*51(8), 56–59.
[6] Okenyi, P.O., and Owens, T.J. (2007). On the Anatomy of Human Hacking. *Information Systems Security* 16, 302–314.
[7] Hobbes, T. (2010). *Leviathan: Or the Matter, Forme, and Power of a Common-Wealth Ecclesiasticall and Civill.* Cambridge University Press, Cambridge, UK.
[8] Tyler, T. (2006). *Why People Obey the Law*. Princeton University Press, Princeton, USA.
[9] IETF. (2014). *On Consensus and Humming in the IETF.* https://tools.ietf.org/html/rfc7282

1

Identity and the Internet

Akram Alfawakheeri

"Some of the most valuable possessions of a refugee in flight are their personal and legal documents. Not only do they prove their citizenship, but they also serve as evidence of their lives before fleeing their country of origin. Keeping these documents safe was therefore an important concern of all refugees. Some refugees sent key documents to family and friends already living in the host country. Others stored these on their smartphone and if possible on cloud services. The latter was considered the safest option given the numerous challenges on the road."

1.1 Introduction

We frequently hear the term "digital identity" under a variety of contexts, especially with the increasing interest in identity as a concept, across a broad spectrum of issues.

Identity politics, digital identity, and self-sovereign identity, all these terms or at least one of them must sound familiar. Despite the enormous and wide-ranging interest in "identity," the concept itself remains something of a puzzle. The meaning of "identity" as it is currently used is not well captured by any given definition. This observation is possibly due to the difficulty in providing an accurate definition that is able to encapsulate the wide range or its present meanings.

The result of a web search on "what is digital identity?" will yield numerous results each with a different definition of the term.

For example, the term can be true for a verified online bank account, and as much valid as to point to the set of information used to identify an external agent for a certain network, that external agent can be any electronic device rather than a human being, for example, as in the Internet of Things.

DOI: 10.1201/9781003339724-3

We believe that before referring to identity, digital identity, the internet, and future visions, it is essential to articulate a common definition of "identity." Moreover, connecting "identity" to the Humanized Internet vision is core to this discussion.

1.2 Identity As We Know It

Why is it necessary to identify ourselves and moreover identify others? One simply needs to know that you are you, you are a part of some society, you go to work, you pay the bills and that is all, correct? Our ancestors pursued those thoughts, to comprehend the concept of being, as a part of a journey to understand the world, its surroundings, but most importantly to understand how we as humans should live our lives and construct societies.

For any social interaction to take place, individuals need to be known and identified. If people do not possess a distinct persistent identity, then promises, obligations, and responsibilities have no meaning.

A phenomenon has emerged in the past couple of decades prompting us to redefine our understanding of identity via technological advancements in the form of artificial intelligence (AI). The use of AI, for example, is raising controversial questions about ethics, about how it can be implemented for better or worse to clarify what makes us human. In the form of AI, and machine learning, patterns of human behavior can be extrapolated, but how accurate? The internet has provided us with various communications media like never before; we can now interact with chatbots in various foreign languages. However, does a chatbot define you or I? Put another way, does a chatbot possess an identity?

To the question "What is identity?" one might respond with the common understanding of identity in the form of "Who are you?" Although this answer might be simple, it will depend heavily on the context; the answer might be "an Australian," "a basketball player," "an accountant," "a grandfather"; or in other cases, the answer might be a document or an email address/password combination.

Two things to notice from this simple question: first, that one can possess multiple identities, and second, each of these identities falls in a different category of the definition for the term "identity"; "an Australian" has a political meaning usually explaining that a person has an Australian national identity, while "a Catholic" refers to the social meaning of identity as belonging to a certain religious group.

However, when referring to an Internet Protocol or IP address, a private encrypted key, and email address, the meaning falls into a historically new definition of identity, digital identity.

Therefore, identity in its present manifestation encompasses multiple dimensions: from traditional personal and social categorizations, to emerging forms of digital representation, all of which contribute to how individuals are recognized and interact within modern society.

Let's get a bit more concrete and discuss an identity example from the personal experience of this book's co-author.

1.3 The Invisible Population: A Personal Story

I would assume in the general notion of the modern meaning of identity; if you ask someone "What is identity?" most people would think of some sort of identification document; maybe the first thing that could cross their mind is a government issued ID card, a passport, or a driver's license. These examples are identity proofs.

Imagine for a moment, that you do not possess a birth certificate, national identity card, and so on.

Now, try to answer the same question again: "What is Identity?"

Such a question would not cause you to question yourself as a person, but it certainly makes it hard for you to prove who you are.

Think about it, a man with no identity card of any sort "does not exist" according to our current systems. I had to think about this issue the hard way; I unfortunately did not just take a moment and close my eyes to imagine what life could look like without having an identification document. It happened when my status suddenly changed from a foreign student enjoying the charming Greek climate to, well … no one.

I lost my passport while I was studying in Greece which had my resident permit sticker on it. I could not renew my resident permit without a passport, and I could not obtain a new passport, because there was no embassy of my home country in Greece due to a political conflict.

I spent weeks running in circles, and I still remember the moment of confusion when, after a long struggle, I saw my data on the screen of the Foreign Affairs office clerk. The records confirmed that I had valid residency, yet to the authorities, I was 'illegal'—simply because I lacked a specific document that only they could issue.

I existed in the system, yet at the same time, I didn't. As paradoxical as that sounds, I walked out of the building that day with my world turned upside down.

So, I was in the system, but also not really in the system, as confusing as this sentence sounds, I left the building that day with my world turned upside down.

To make matters worse, I couldn't book a flight, enroll in a new semester, or even renew my sublet contract. As the days passed, I realized I was completely unable to take any action—no matter how ordinary.

I began questioning the entire system, something I had always taken for granted, something I never thought could fail me. When I had heard the word 'stateless' before, I had only associated it with nomadic communities or marginalized groups. But now, I was forced to confront a broader definition—one that I had unknowingly become a part of. That was the beginning of an unexpected journey—one that changed my life forever, but also transformed the way I see the world. This experience ignited a deep interest in understanding identity, identity systems, and their flaws.

Though what happened to me was difficult, I now see it as a small hardship compared to the greater realities so many people face every day.

I felt like Viktor Navorski, the main character of Steven Spielberg's "The Terminal" movie. Viktor is a man who got caught up in international politics, shortly after he arrived in New York when he discovered that his passport was no longer valid, resulting in him being trapped at JFK airport, although I was trapped in a country not in an airport. It was then I learned that the movie was a romanticized story of Mehran Karimi Nasseri, for whom a small corner of the international transit zone in Charles De Gaulle Airport's Terminal One was the "home" for 18 years (August 1988 until July 2006).

In 2018, a similar story occurred again, this time in Kuala Lumpur International Airport to Hassan Al Kontar, a Syrian citizen, in 2018 who was also trapped in a situation of international politics and a bureaucratic limbo that stretched out for 9 months.

A much crueler story was illustrated in the movie, the story of Meena Haseena, a story of how an invisible child can be an easy victim of human trafficking and sex slavery (https://www.meenathemovie.com/).

Examples are too many to cover, and there are more horrific stories out there. But the most horrific fact is that this is the everyday reality of over 1 billion people globally who do not have any official identity.

From this group of individuals, 81% reside in South Asia and sub-Saharan Africa; 47% are below the national ID age, with women and the poorest who risk being left behind due to the lack of an ID and further potential victims of human trafficking.

Losing a passport probably doesn't make you question who you are, but it certainly strips you of the ability to prove it. But what if you never had an official ID or even a birth certificate—and couldn't simply apply for one?

Perhaps the country where you were born and raised refuses to recognize people from your ethnic group as citizens, like the Rohingya population.

Or maybe you had to flee your homeland due to war or a natural disaster, making it impossible to return and reissue your documents. Or what if your country simply ceased to exist? In those cases, what defines you? What gives a piece of paper such immense power over your life?

How did we arrive at a reality where your worth is determined not by who you are, but by the documents you hold? Should it all come down to the randomness of birth? Why must one prove their existence just to be acknowledged? And by whose authority?

Have you ever stood before an immigration officer, waiting to present a supplementary residency card—required for re-entry into the country where you legally reside—only to realize you forgot it? Even though the officer can see your records in the system, they still deliberate whether to let you in. In that moment, your fate is determined not by your identity, but by a document.

Now imagine that being your permanent reality—the loss of dignity that so many people endure today.

Or, perhaps you had to flee your country because of a natural disaster or a war and you cannot go back to have your documents re-issued. Or your country ceased to exist. In those cases, what identifies you? What gives such a piece of paper the magical power over your life? How did we arrive at such a political and social situation where your worth is determined not by who you are but by the type of paper you hold? Should it all come down to the pure chance of where you were born? Why does one need to prove their existence to actually exist and by which authority? Have you ever stood before an immigration officer waiting for you to produce a supplement residency card to your passport required for re-entry into the country of your residence and you had forgotten to take the card with you when traveling, in spite of the fact that the immigration officer "sees you in the country of residence system" and must deliberate whether or not you should be allowed to re-enter the country? Your fate is suddenly decided by possession or non-possession of a document that accords you certain rights, like to live and thrive in a country. Without it, you experience the loss of dignity that so many people are enduring today.

As written in Article One of the Universal Declaration of Human Rights, "All human beings are born free and equal in dignity and rights."

With forced migration due to war, conflicts, poverty, and climate change, the number is getting bigger, 1 billion people lack the most basic human right, the right of personhood, and without it other human rights become out of reach, like the right to belong to a country, the right to learn, work, and the right of freedom of movement. Basic human rights and needs are out of reach without an officially recognized identity.

Another staggering statistic I've encountered is the fact that over 5 billion people are active on social media. The concern isn't just the immense traffic these platforms generate, but rather the business model that thrives on people's behaviors, data, activities, and online presence—in other words, their digital identity. Whether you actively engage or not, you have an online presence that goes far beyond simply having an email address or a social media profile. It encompasses your behaviors, shopping habits, financial standing, interests, and even the results of personality and psychological assessments.

Big companies like Google, Netflix, and Amazon use ever-evolving algorithms to direct your consumer behavior. Algorithms track your general online habits even outside those websites, to calculate what to sell you. You are a product by definition.

Perhaps it is crucial to understand the principles by which these algorithms operate and the data these companies collect about you and your identity in order to fully grasp the influence they have on your life.

We now live in a reality where some people lack any form of official identity, while others are 'over-identified' as digital products. Traditionally, aspects like gender, age, and profession have been key components of identity, while personal preferences—such as favorite childhood shows, music taste, or outfit colors—were not. However, with the rise of the internet, no data is considered 'unimportant.' People are now 'over-identified,' with everything from political beliefs to food preferences, religious views to video game activities, being collected, analyzed, and monetized. We now live in a reality where some people lack any form of official identity, while others are 'over-identified' as digital products. Traditionally, aspects like gender, age, and profession have been key components of identity, while personal preferences—such as favorite childhood shows, music taste, or outfit colors—were not. However, with the rise of the internet, no data is considered 'unimportant.' People are now 'over-identified,' with everything from political beliefs to food preferences, religious views to video game activities, being collected, analyzed, and monetized.

In the next chapter, we will talk more about what we mean by this term, we will also talk more about digital identities, the internet, and what are the problems regarding our rights and privacy we are facing.

But before that, if we need to understand the problem, we need to see the bigger picture, and we need to understand how our identity system has changed and developed over time.

So, let's dive into a brief Identity history.

1.4 A Brief History of Identity

Before World War I (WWI), most people did not possess or need an identity document or passport.

The cave paintings of prehistoric people around the world are dated as far back as 40,000 years ago and are believed to be the oldest traces of identity, maybe at least in the meaning of the human desire to leave a unique mark among others and to be remembered. [Marchant, Jo (2016)]

However, identification by fingerprints that is still used today may go back as old Babylon or ancient China where these societies were the first to fully use biometrics as identifiers. From circa 206 BC to 589 CE, fingerprints were used as identifiers in all sorts of situations. A combination of fingerprints, body measurements, footprints, and handprints were recorded on paper or silk. This combination would be used on official documents of all kinds, including engagements, divorces, deeds, records of indenture, and even army records. There is evidence that fingerprints were used in crime scene investigations during that time as well.

In more recent history, Sir William James Herschel and the author of the book "The Origin of Fingerprinting" might be considered the first European to use fingerprints for administrative purposes while serving in Bengal around 1858.

The first known identity document was a form of passport introduced by King Henry V of England through the Safe Conduct Act of 1414 [Act of Parliament]. Interestingly, this document was issued by the king to anyone, regardless of whether they were English or not, and this practice continued until the 1850s.

Before World War I, most people did not possess or need an identity document or passport, and the current form of photo ID became widely used only until the early 20th century. A standard idea of a worldwide passport was born from the aftermath of WWI, particularly after the *Paris Peace Conference*, and as a step taken by the League of Nations toward maintaining peace.

At the end of WW1, the whole world was a mess, many governments were toppled or fell, and borders were changed and redrawn, resulting in millions of refugees, ex-prisoners, and stateless people. In 1922, the first refugee travel document was issued by the *League of Nations* as well. This document is known as the *Nansen Passport.* The early versions of the passports were signed and had a photograph, as well as personal details like physical details such as marks and shape of face. Some critics of early travelers found this dehumanizing.

What appeared to be a measure to maintain peace, provide freedom, and assist refugees could arguably be seen as a political opportunity to exert further control over movement and immigration— a pattern that became evident in the restrictive migration policies adopted by many nations.

History, of course, offers even more powerful examples, particularly during World War II, when identity labeling reached extreme levels.

It seemed that people struggled with the implicit labeling, packaging associated with the passport concept, even some countries wanted to dispose of the passport concept. However, no one could reverse the idea, no country nor an individual could choose not having a passport.

Current passports have retained the shape of earlier passports. However, passports have evolved to include multiple security features that can count up to more than 80 on some passports, like holograms, watermarks, and barcodes. Other modern passports have an RFID chip nested in them which includes the personal information of the passport holder and other details.

These examples are the so-called "improvements" that have taken place on the passport as an object but not as a concept. We are in the 21st century and witnessing the changes of states, borders, and discrimination against ethnic groups that have resulted in the increased number of stateless people to more than 10 million people. These people are denied passports, thus the freedom of movement or, in some cases, the absence of it. These extremes illustrate how nebulous our notions of citizenship really are.

On the other hand, passports as well as identity or citizenship have become a highly sought-after commodity, and here, we are not referring to a black-market of stolen, well-forged passports and identity cards. There is an entire legal market for citizenship and passports. Many countries offer citizenship as a minimum investment in a country, for example, Cyprus will provide citizenship for $1 million investment in the country. There are even tools on the internet to search for available passport purchases and rank them, having an annual announcement of the "best" and "worse," in addition to luxury services for wealthy individuals who are interested in pursuing a second or even third passport.

Borders are increasingly open to those who can pay, viewing it as an investment in a country. This highlights how identity and citizenship are no longer defined by fixed boundaries but rather as flexible transactions. As Clifford suggests, 'What if identity is conceived not as a boundary to be maintained, but as a nexus of relations and transactions actively engaging a subject?'

This philosophical shift reflects the commodification of citizenship in a globalized world, where relationships and transactions—not static boundaries—determine who belongs and who does not.

In our current reality, citizenship is an accident of birth, defined by borders that were drawn years ago by people who are most probably not from the same region they drew borders for.

The passport that you hold can be a shield and a privilege if it belongs to a wealthy democracy, and it can be a distress and a burden if it belongs to an unfortunate, poor, or politically conflicted country.

The world is changing faster than ever, identity citizenship and passports need to evolve essentially to meet the near conceptional shifting in the core meaning of "you" and "I." Is not about passwords, bank info, social security numbers anymore, it's about behavior "likes," shopping list, personality, and psychological behavior.

1.5 Digital Identity World Order: Toward Self-Sovereign Identity

In the previous section, we discussed the notion of "analog identity" that is associated with paper documents like driving licenses, birth certificates, and passports.

Identity and digital "citizenship" are being developed by social media giants like Alphabet Inc., Meta, and X for example. Connecting the next billion people to the internet may seem noble as a mission, but is it? Or, are you just a product of techno-colonial expansion? – as the assertion, "If you are not paying for it, then you are the product," has been previously made.

These organizations validate who we are and formulate a digital profile of us, a sort of "digital apostle" as a new type of citizen. We share different aspects of ourselves in different situations and for differing reasons. Our identity solutions should be flexible enough to accommodate our multiple selves. Therefore, we must ask: Can a person possess multiple identities?

The obvious response is 'yes,' as we all embody different personae—private, professional, and so on.

Do we want central organizations to provide us with a digital identity, or should we determine our own self-sovereignty? However, the 'use' of identities by nations can be easily abused, and we don't have to look far in history for examples.

We've certainly seen the weaponization of identity, whether through ethnicity or religion—such as the Nazis during World War II. More recently, the challenge to 'sanctuary cities' in the United States has sparked debates around the treatment of undocumented individuals, highlighting the ongoing issues with identity and borders.

On May 17, 2018, one of the authors had an opportunity to participate in a workshop, entitled, "Digital Identity and the Future of Democracy."

As we have stated earlier in this chapter, human rights are universal, at least from a western constitutional perspective. With the increasing instability of states and economies in recent times, resulting in new waves of migration worldwide and the refugee crisis, in particular, human dignity is under threat in various respects. One very basic aspect is the loss of identity and citizenship of the individual frequently caused by the circumstances of migration. Digital identity is foundational to other services and business beyond national borders, including e-Government if it is to be made effective.

It empowers the individual having control over her/his data independently of the nationality and place of residence or business. It can help, first of all, to establish citizenship of the person of his/her state of origin. Without this, the regime of statelessness would apply in certain cases, with the consequence that almost no rights are granted to the person, nor diplomatic protection, etc. In an increasingly mobile world with more and more porous borders, it is essential for everybody to dispose of a reliable tool for proving her identity and professional skills. It can be developed to a digital identity wallet facilitating authentication and verification in international trade, but also the development of new data-based business models. Digital identity, thus, is foundational to social trust and can facilitate mobility and promote business in the digitalized world.

Digital identity can extend beyond its current use. It could serve as a foundation for recognizing an individual's status as a global citizen, independent of nationality, thus allowing participation in global democratic processes and the formation of collective will. This would be essential for establishing a system of regulation to address global challenges. A normative framework for the operation of a globally free internet and communication system is only one of these challenges. Others are climate change, frameworks for sustainable international trade, global financial markets, the fight against terrorism, and organized crime. The Global Risks Report 2018 presented to the Davos World Economic Forum witnesses our challenges ahead. Yet, our present democratic systems at the national and supranational level are unable to meet these challenges which are beyond the reach of national politics. Even international cooperation fails to produce effective regulation as is needed. David Held has given a seminal account of the deficiencies of the international system already in the nineties of last century. As Jürgen Habermas explains, the establishment of regulatory power beyond the state, globally, is not a challenge of, but in itself a requirement of democracy.

Digital identity and, based upon it, global citizenship as a door to participation in democratic decision-making processes at the global level do not substitute the citizenship of a state, nor does an institutional setting for global regulation substitute the state. It is additional and complementary to national citizenship and the state, although the system may offer the establishment of digital identity, registration, and the operation of global citizenship independently of a particular nation state.

It is the basis for conceptualizing democratically legitimized regulation as needed to meet global challenges effectively. Global constitutionalism built upon these insights could become a realistic utopia if coupled with the new opportunities offered by the internet. The question, thus, has been to explore how technologies make democratic processes effective at the global level.

The opportunity for digital self-sovereignty is that you become the center of the universe in determining what you want to share about you and with what method. This is NOT about global citizenship, no not at all. Could the individual enabled with technology have digital safe boxes for his/her documentation that can be digitally notarized in near real time? Rather than "centralization of data" which is often hacked and/or abused, why not think about a hybrid?

1.6 Summary

While identity is complex, digital identity and the internet possess multiple attributes such as the notion of digital citizenship, digital identity abuse, and the centralization of digital data by organizations in the form of metadata where algorithms profile you. As previously discussed in the book Introduction, the Humanized Internet challenges the digital status quo where questions of ethics, cognitive bias recognition, and the opportunities for Self-Sovereign Identity can be the basis of a 21st-century internet. Imagine the human as an active player and not machines. In November 2019, Sir Tim Berners-Lee articulated Nine Principles for Contract of the Web engaging the very internet companies which have lacked a self-governance framework. Can such a model succeed beyond the call for manifestos?

In Chapter two, we discuss the development of a human participatory internet and the enabling technologies where self-sovereign identity and the fair treatment of data can be foundational. The topic is only the beginning and a dystopian future of algorithmic control is what we would like to avoid. Is it too late?

References

[1] Michael J. Casey and Paul Vigna (2018). *The Truth Machine, The Blockchain and Future of Everything*. HarperCollins Publishers, New York, NY.

[2] https://phys.org/news/2018-04-social-media-platforms.html and https://www.forbes.com/sites/marketshare/2012/03/05/if-youre-not-paying-for-it-you-become-the-product/#92e862d5d6ee

[3] Ann Greenberg and Monique J. Morrow (2017). *SXSW 2017 Panel Picker Proposal*. https://www.slideshare.net/anngreenberg123/identity-are-you-in-control-108653672

[4] Edwin Black (2009). *Nazi Nexus: America's Corporate Connections to Hitler's Holocaust*. Publisher: Dialog Press.

[5] Jürgen Habermas (2012). *The Crisis of the European Union: A Response*. Polity Press, Cambridge, UK.

[6] https://www.jura.uni-bremen.de/uploads/Viellechner/HIIG_Workshop.pdf

[7] https://www.weforum.org/press/2018/01/digital-identity-why-it-matters-and-why-it-s-important-we-get-it-right/

[8] https://www.theverge.com/2019/11/25/20981502/contract-for-the-web-tim-berners-lee-google-facebook-principles-techlash

[9] http://id4d.worldbank.org/global-dataset

[10] Marchant, Jo (2016). Journey to the Oldest Cave Paintings in the World. *Smithsonian Magazine*. http://www.smithsonianmag.com/history/journey-oldest-cave-paintings-world-180957685/

[11] Alencar, A., Kondova, K., & Ribbens, W. (2019). The smartphone as a lifeline: an exploration of refugees' use of mobile communication technologies during their flight. *Media, Culture & Society*, 41(6), 828–844. https://doi.org/10.1177/0163443718813486

[12] http://id4d.worldbank.org/global-dataset

[13] Daluz, Hillary Moses. (2014). *Fundamentals of Fingerprint Analysis*. CRC Press, Boca Raton.

[14] Sengoopta, Chandak. (2003). *Imprint of the Raj: How Fingerprinting Was Born in Colonial India*

[15] Kavanagh, Jennifer. (2010). *Journey Home: An exploration of our inner and outer identity.* https://web.archive.org/web/20061125184446/http://www.ukpa.gov.uk/general_history_early.asp

[16] https://www.canada.ca/en/immigration-refugees-citizenship/services/canadians/celebrate-being-canadian/teachers-corner/history-passports.html; http://www.refworld.org/docid/3dd8b4864.html

[17] Emergency Quota Act of 1921 and later, the Immigration Act of 1924.
[18] Immigration Act of 1924 (Johnson–Reed Act) P.L. 68–139; 43 Stat. 153
[19] Mark B. Salter (2003). *Rights of Passage: The Passport in International Relations*. http://uk.businessinsider.com/when-governments-wanted-to-get-rid-of-passports-2016-10?r=US&IR=T
[20] https://www.theguardian.com/travel/2006/nov/17/travelnews
[21] http://www.unhcr.org/stateless-people.html
[22] https://www.nytimes.com/2015/05/01/business/dealbook/malta-offers-citizenship-and-all-its-perks-for-a-price.html; http://www.moi.gov.cy/moi/moi.nsf/All/36DB428D50A58C00C2257C1B00218CAB]
[23] Clifford, James. 1988. The Predicament of Culture: Twentieth-Century Ethnography, Literature, and Art. Cambridge, MA: Harvard University Press.
[24] https://edition.cnn.com/interactitove/2017/01/politics/immigration-ban-stories/
[25] https://www.theverge.com/2019/11/25/20981502/contract-for-the-web-tim-berners-lee-google-facebook-principles-techlash

2

Evolution of the Internet and New Technological Developments

"The Internet is the first thing that humanity has built that humanity doesn't understand, the largest experiment in anarchy that we have ever had."

Eric Schmidt

2.1 Introduction

Continuing our journey that has taken us to the Humanized Internet with Digital Identity and the related personal narratives as foundational. The COVID-19 pandemic posed the question as to what is private, non-intrusive and safe. New contact tracing applications were developed to facilitate the tracking of an infected person with the promise that no personal data will be violated, which is used without the explicit agreement of the person using the application.

The polarity between privacy, safety and self-sovereign identity continues. After all, are you not surveilled when using a global positioning system (GPS) on your smartphone? What is perhaps more disturbing is the isolation that has resulted from COVID-19-related interventions, whether from the loss of jobs or from economic shutdowns, e.g., restaurants, tourism, and so on. As an observer, one could pose the question of whether or not the loss of liberty and associated isolationism has allowed more time for the internet to be used as a platform for discussions related to trust (or lack of), conspiracy theories (Q-Anon as an example). Perhaps the sense that one is losing control exacerbates this narrative. What is true, who defines truth and by which standards is contextual. Deep fakes are a reality; and Pandora's box of platforms seemingly proliferates a sense of belonging or an amplification of siloed narratives, begs the question of what is the governance model for these platforms. Who decides the usage of these platforms and for what purpose? These social media platforms have been self-regulating and therein lies the dilemma. Which entities

DOI: 10.1201/9781003339724-4

monitor social media and by which rules? This issue has become political hence the requirement for a shared global agreement of what comprises, "Do No Harm" and what specifically are its attributes. The internet has been evolving and is so embedded in our daily life that articulating responsible use across this vast ecosystem is complex, and perhaps an opportunity.

There is no doubt that humanity must benefit from the solutions that can be created using technology. Technologies such as AI, machine learning, blockchain, and quantum computing serve as enablers for innovation; and, the promise is bright for our society.

With smartphones, smart homes and smart "x" we pulsate information daily. The Internet of Things exacerbates the risk of data leakage that may result in data misuse, user profiling and user traceability. Data privacy laws focus on unlawful data processing in the form of data audit enforcements and impactful corporate penalties. However, with data security, there is a correlation to safety in the prevention of harm to the individual and harm to an organization, the latter often in the form of ransomware. At a macro level, there is a growing threat to the weaponization of data such that "live by the digit, die by the digit" will be embedded ominously in our society.

It is worth exploring the evolution of the internet as we know it today and the emerging technologies that are defining new layers and functions that challenge current social systems. This challenge is manifested in "control by a few large organizations" that have resulted in productizing individuals without their knowledge in the form of metadata. How did the internet evolve from a government experiment to a full internet economy where self-governance has not been forthcoming? Simply put, free is not free and perhaps we must pause and contemplate what control means to the people who use the internet? Heads of government communicate in real time over social media. No doubt that the internet has been a platform for communications and knowledge exchange, but is there a risk that knowledge exchange may be a proliferation of fake news or perhaps what is fake may be a filtered truth embedded in the internet. If we agree that there is a trust deficit, how do we address it and in what format? Regulators have been careful not to over-rotate on regulation as to allow the internet to be market-driven. Microsoft's Brad Smith has called for Big Tech to be regulated as self-regulation has simply failed.

Microsoft and Apple have asserted that privacy is a fundamental human right.

Facebook now Meta has fumbled throughout these discussions with missteps that only erode the trust component of this discussion.

Further, you as a consumer do not need to be logged in to be tracked by Google.

The narrative here is not so black and white. Faulting these companies because of their successful business models would be easy. What is essential is whether or not privacy is dead and/or whether we can control our multiple personae over the internet. The discussion of self-sovereign identity (SSI) and the internet is key to understanding your role in selective disclosure.

Of course, the question of ethics, technology and its intentional use is still pivotal now more than ever.

Whether artificial intelligence (AI) or brain interfaces to the internet, technologies are developing rapidly and with it, the potential for greater abuse. For example, exploration in telepathy with brain interfaces is occurring now.

"I know what you are thinking. You cannot hide now...."

Hello, new world! Can you imagine the potential for abuse or will it matter? Perhaps we will need to have a privacy button developed as an option to allow for some level of randomness and serendipity. Let us take this notion of "decentralized" and its execution a bit further. Can "things" and "people" operate autonomously where "authority" is defined within these constructs?

The decentralized paradigm is a work in progress shaping a "new internet," that indeed embeds a notion of shared ethics and where data is under the purview of the individual or even more provocatively, the evolving organism or thing. The brain as an interface to the internet has been under development for the past several years to foster now, the "Brainternet." However, with the development of portable MRIs, could we not think about the potential of "telepathic exchanges" or brain-to-brain messaging, and if so, why would we need internet overall? In essence, we BECOME the internet.

Further, we can now add our own "serendipity on-off buttons" and choose to share WHAT we wish to share and with WHOM we wish to share. However, will I be able to fend off a new kind of denial-of-service attack on my brain? What is the potential for abuse in this "brave new world" should our brains become "co-opted" by a few miscreant actors too far off in the future? We must note that humans are the creators of these technologies and algorithms. However, when there is no embedded safe use policy, things can and do go wrong. It is important for you as our reader to decide.

How we got here and where we are going with the internet, is an opportunity for us to pause and truly reflect the internet's impact on our society. Let us take a look briefly at how the internet evolved to where we are today.

2.2 Brief History of the Internet

In 1969 DARPA funded ARPANET (Advanced Research Projects Agency Network) which formed the base and foundation for the internet – packet switch network and implementation of TCP/IP. The network allowed general communications of users with each other on a single network. While the internet was not invented at this time, by 1971 the first email was sent over the ARPANET and by the mid-1970s email was established. ARPANET adopted TCP/IP on January 1, 1983, where the "network of networks" expanded and became the modern internet. In 1990 the online world was born when computer scientist Tim Berners-Lee invented the World Wide Web – a method of accessing data online in the form of websites and hyperlinks. By August 6, 1991, he had built the first website while at CERN. The Web's capabilities truly catapulted the internet with the public masses when the technology was made available royalty-free by 1993 which ultimately helped expand the wealth of information – accessed mostly via dial-up modems. *This was labeled later as Web 1.0.*

The sequence of history paved the way for the birth of new internet-based companies where in the 1990s some of today's online giants such as Yahoo which was launched, Amazon was Incorporated, eBay was founded and Google was born. The public embraced these new digital companies while at the same time, entrepreneurs and investors realized a new and much faster and cheaper ability to develop novel online companies through new emerging technologies such as cloud computing.

In the 2000s, when broadband became mainstream, the web expanded with the creation of Wikipedia in 2001 followed shortly when the first social network website was created – Meta was born in 2004 and became mainstream. Around the same period "Cloud Computing" came into existence with Amazon announcing in 2006 the creation of Amazon Web Services and introduced its Elastic Compute Cloud (EC2). Following it Google released in 2008 the Google App Engine with other vendors such as Microsoft and Rackspace creating their own public cloud versions to allow anyone in the world to access computing and development resources at any time, at a fraction of the cost of traditional on-premise methods. Cloud computing allowed literally anyone to create new companies online at speeds and scales as never seen in history before. These developments enhanced the opportunity for new digital companies to emerge with the likes of Linkedin, Reddit, X, Google+, Instagram, Quora, Dropbox, Skype, and many others which were launched. *Welcome Web 2.0!*

The wheels for the next-generation web, *Web 3.0*, have already been set in motion. Unlike the more centralized and much controlled Web 1.0 and Web 2.0, Web 3.0 calls for a more open source, neural, borderless, open innovation, open access, censorship resistant, and more secure and privacy-enhancing web that is a Humanized Internet for the people and by the people where people are in control of their own data, identity, and destiny.

It should be stated that in parallel to these internet developments has been the growth of the mobile phone. The mobile phone is in essence an extension of ourselves by which we interact with the internet. Imagine the panic that some of us have experienced at the loss and/or misplacement of our mobile phone. The growth of the mobile phone has in fact fueled the internet itself. Perhaps better stated, these two phenomena are intertwined. It is no wonder that the discussion of "Internet Democratization" is rooted in providing internet access to everyone globally. The reasoning is access to education, healthcare, and the eventual digitalization of the workforce there in fostering national transformation programs that could result in economic growth and perhaps an equalization of the north-south divide that we observe today.

Moving forward, as Web 3.0 evolves, inclusion and trust will be core elements and differentiators for this multi-stakeholder dialogue.

The "Contract for the Web" is a good start; however, it includes the very companies that have been called out for their business centralization models and monetization of consumer data. How will the "Contract for the Web" be enforced and by which entities? Will there be a governance model where the consumer is measuring enforceability? We cannot say for certain, but perhaps pivoting is an opportunity for companies and organizations toward a larger vision that is truly for everyone.

As we turn to 2023 and beyond, more than 4.5 billion people of a population of over 7 billion individuals are connected to the internet. The use of the mobile phone in Africa for example, has accelerated the internet adoption pace and leapfrogged traditional models thereby stimulating innovation, for example in mobile payment schemes like Kenya's M-Pesa, is the tip of the iceberg of what is possible. However, there is a need to be cognizant of "techno colonialism" by global technology companies in Africa and in other geographies. The desirable model could well be developed in Africa, for Africa and adopted by the world. Why not? The internet is evolving and with it hubs of opportunity that are driven elsewhere. Indeed the discussion is contextual and only presents opportunities for culturally inclusive digital transformation.

The Internet of Trust is thematic as we have noted. For example "In x we Trust and Why "provokes various discussions as to the polarities we are experiencing between, privacy, perceived loss of control, security and ethics." Rather than deprecating to what could be a dystopian narrative, there are emerging technologies such as self-sovereign identity or SSI that have the potential to empower the individual to assert control of his/her identity selectively.

The Internet of Trust and self-sovereign identity are no longer ethereal monikers but very real. The thesis is that YOU are the center of this data and identity universe that includes of course the internet. Let us double-click further the implications of the Internet of Trust and self-sovereign identity to you.

2.3 The Internet of Trust and Self-Sovereign Identity

As the old adage goes, "Trust takes years to build, seconds to break and forever to repair." The Internet of Trust is core when articulating the values of Humanized Internet. When your data is leaked out by accident, hacked in a centralized environment or monetized without your permission, your trust is violated. Who do you trust and why? Standardizing on "Fair Data" use like "Fair Trade" may be interesting insofar that the data sources are known by the individuals who may represent these sources. Specifically, what data is trusted, by whom and how these data are treated and for what purpose is critical to this topic of "Fair Data and the Internet of Trust."

Moreover, the World Economic Forum has been focused on this very issue of data policy with the formation of a Data Policy Council that includes academia, companies, and organizations addressing the topic of data policy that is complex and contextual. One of the co-authors of this book participated in the World Economic Forum Data Policy Council.

There is an intersectionality between data privacy, cybersecurity, data governance, and data standards, which are indeed common themes that are interwoven throughout this book. As an observation, there is a distinction between "data about me" and what is "my data." At a macro level, "data about me" may be that which is derived about you digitally over the internet, the digital forensic prints that you leave every time you use the internet publicly. These digital forensics are forever and removing them has its own set of challenges. "My Data" may have inferences to Personal Identifiable Information (PII) such as date of birth, social security number, address, and so on. Understanding which entities other than you possess such data and implications of leakage is critical to the privacy discussion. However, when referring to data monetization and algorithmic profiling, the context is usually under "data about me." Though leakage of "my data" can result in identity theft, one can still infer the identity of an individual in spite of data anonymization and metadata techniques, some of the research found in *differential privacy* and described well in The Ethical Algorithm, The

Science of Socially Aware Algorithm Design, by Michael Kearns and Aaron Roth. The point is that understanding this difference assists in framing the discussion clearly.

Moving to the self-sovereign identity of SSI is really where the consumer is in the middle of this Copernican Internet universe.

The attributes of SSI are:

- **Minimalistic**: Discloses the least amount of data or what is commonly known as "selective disclosure."
- **Security**: Privacy and confidentiality are basic tenets.
- **Control**: The consumer controls his/her/ personae or identification.
- **Consent**: Identification is used only if the consumer consents to its use.
- **Portability**: There is an interoperability aspect that is implied.
- **Persistent**: The identification can never be taken away from the consumer.
- **Resistant**: There is resistance to censorship and identification deletion.

From the overview of self-sovereign identity published in November 18 2020, Internet Policy Review Journal, the authors assert:

> "Self-sovereign identity (SSI) is rooted in the belief that individuals have the right to an identity independent of reliance on a third-party identity provider, such as the state or any other central authority. Its implementation requires the development of technical standards, as well as socio-political adaptations rooted in legal amendments in order to be successful. Overall, *SSI is implemented as blockchain adjacent, but not blockchain-dependent identity management systems, which are guided by the fundamental principle of user-centric design, using technical standards that enable user-generated and user-controlled decentralized identifiers, associated credentials, and attestations*."

In 2019, the European Union published a paper positioning SSI implementation with Electronic Identification, Authentication and Trust Services or eIDAS.

Open Source technology are enablers for SSI like the Linux Foundation's Hyperledger family specifically, Indy, Ursa, and Aries.

W3C organization has been focused on developing SSI use cases specific to credentialing.

This topic may sound nebulous to the reader, but think about it for a moment. Would you like to control selectively which entity should have access to your data and for what purpose? The key word is "selective disclosure." Let us take health data as an example. While regulations in various countries such as General Data Protection Regulation (GDPR) in the European Union or the Health Insurance Portability and Accountability Act (HIPAA) in the United States or the California Consumer Privacy Act (CCPA) in the US state of California, to name a few examples, there are strict regulations as to data handling. However, are you engaged and active in these discussions? The challenge is that these regulations are contextual and siloed under a geo-political umbrella.

Even if your data is anonymized where the result is metadata, you are still not in control. Again, which entities are anonymizing your data and for what purpose? It is far too easy to assert that privacy is dead and selective control is no longer valid. Moreover, I want to know how data about me and/or my data is being used by which entities and yes, for what purpose. SSI has been developing for the past several years with use cases including supply chain management at the industry level. The decentralization and use of decentralized identifiers by credentialing organizations such as universities or government entities is low-hanging fruit.

Certainly establishing a data governance framework that includes recommendations for industry and citizens/consumers may be considered. Together with the Internet of Trust and self-sovereign identity, the internet is evolving to the formation of a trust layer where policies of safety, privacy, data governance and identity are associated with selected disclosure. The core action here is to ponder the ramifications of selective disclosure and perhaps the opportunity to execute it.

What about the internet and ethics? In fact, whose ethics? In the next section, we will examine the theme of the internet and ethics. Ethical machines will be discussed in Chapter 8.

2.4 Internet and Ethics

Sir Tim Berners-Lee has proposed the creation of "Personal Online Data Stores: or PODs" that provide individuals with the capability to manage their own data.

The premise is that the social media platform companies are not only "siloed" but that they enable the very surveillance that transgresses individual privacy and therefore require a new definition of a shared internet ethics vision.

In fact one of the value-added propositions for the Humanized Internet has been the creation of "Digital Lock Boxes" that would be managed by the individuals. The Humanized Internet Digital Lock Box vision has been articulated in a few public events these past several years.

The Architecture of the Humanized Internet Trusted Digital Lockbox ID is built around various building blocks that comprise the entire solution with the ultimate goal of creating a new environment and values such as promotion of well-being, increased human capability, liberties, and a converged and hyper-connected world.

Some key characteristics of the Trusted Digital Lockbox ID (TDLID) are:

- Decentralized, autonomous, dynamic, human-centric and person controlled, secure, safe while trusting identity and associated at the core.
- Giving more "power" back to the user/person/people and their own identity and data.
- "Co-create" identity and data ability with existing issuers of personal data generators (identity truth and trust fabric).
- TDLID: Uniquely recognizable characteristic of personal data that forms the core identity/entity. Private, secure, and not used over the internet.
- Derived TDLID: Crypto function that uniquely identifies a person or entity. Private, immutable, and not to be used for transactions over internet.
- Shadow TDLID: Various lockboxes (life, work, social, etc.) that are derived from the core TDLID attributes and characteristics.
- Transaction TDLID: Used for transactions and derived from TDLID. Can be blacklisted/removed/temporarily and regenerated from core TDLID.

Figure 2.1 summarizes the components of the "Humanized Internet Digital Lockbox."

Make no mistake, this is a very hard problem to solve and it is satisfying to witness Sir Tim Berners Lee and his team tackle the issue.

In double-clicking on the Internet and Ethics, the propensity for abuse has been top of mind for consumers. In fact, there has been quite a bit of discussion on social media data minimization and which messaging platform

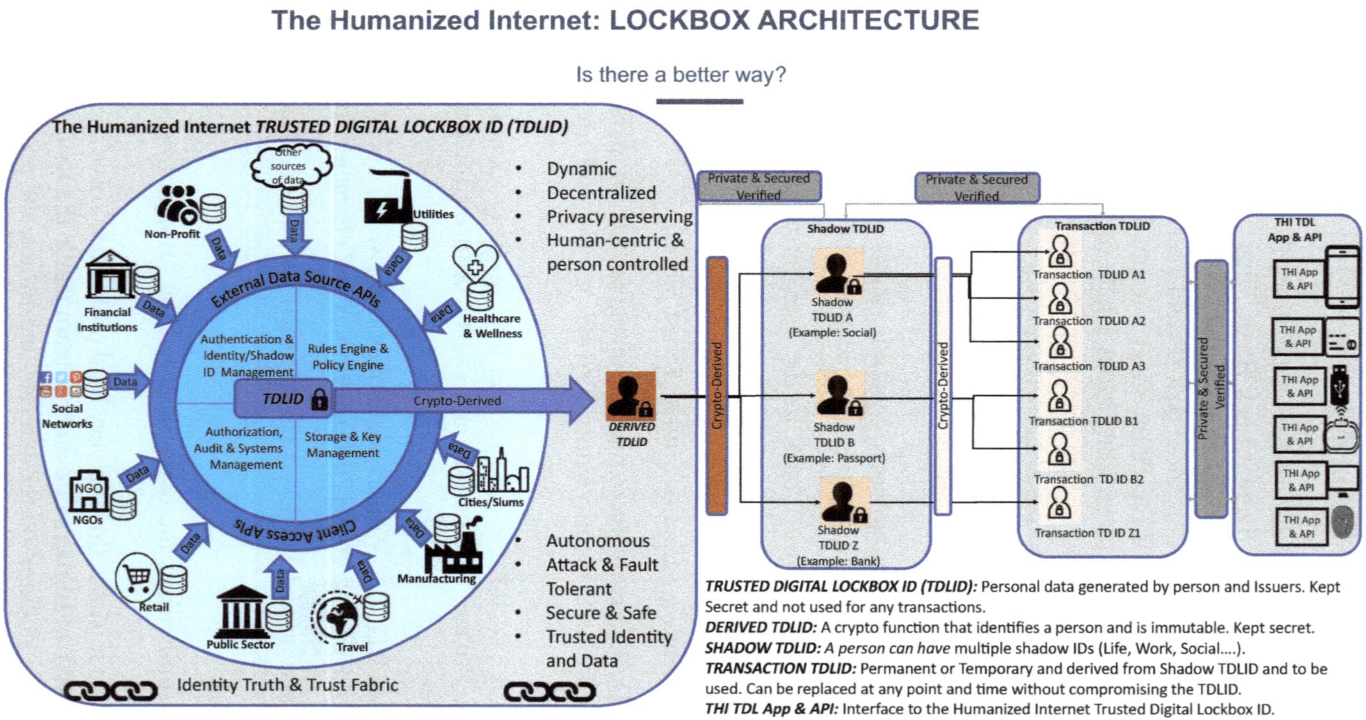

Figure 2.1 The Humanized Internet: Personal Lockbox Architecture.

is "safer," e.g., Signal, Telegram, and so on. Meta has been on the receiving end of this information backlash specific to WhatsApp flows directly to the Meta platform. In spite of capacity building as to what data is captured for which purpose, the question is who do you trust and why? For example, in their book, The Ethical Algorithm, Michael Kearns and Aaron Roth assert, "In many ways deciding which trust model you want is more important than setting the quantitative privacy parameter of your algorithm. But the two decisions cannot be made in isolation." The Internet and Ethics. Embedding ethics in the internet such that we avoid cognitive bias will have its challenges but certainly, there is an opportunity to consider the possibilities. A shared Internet Ethics vision that is multi-stakeholder is certainly a first step. Understanding the triggers for a shared Internet Ethics vision is critical toward developing an articulation of the vision.

2.5 Chapter Summary

Connecting the dots from a personal story involving identity in Chapter 1, we have extended the discussion to the evolution of the internet and implications of technology to our social structure in Chapter 2. Amplifying the polarities between data abuse, privacy, security, data governance and the intersectionality with self-sovereign identity and ethics correlated to the evolution of the internet, the common theme is contextual and responsible use of the internet that engages multi-stakeholders most importantly, putting the individual in the center of this narrative. In Part II, Chapter 3 we will discuss the emergence of new organizations that are coalescing around some of these topics. In Chapter 4, Societal Vulnerabilities and Implications Toward Tokenization of the Self, we propose a new social contract.

References

[1] https://www.npr.org/2019/09/13/760478177/microsoft-president-democracy-is-at-stake-regulate-big-tech]
[2] https://www.bag.admin.ch/bag/en/home/krankheiten/ausbrueche-epidemien-pandemien/aktuelle-ausbrueche-epidemien/novel-cov/swisscovid-app-und-contact-tracing.html
[3] https://www.eweek.com/security/microsoft-ceo-says-privacy-a-basic-human-right-calls-for-global-gdpr
[4] https://www.fastcompany.com/90421349/one-congresswoman-perfectly-summed-up-facebooks-hypocritical-stance-on-privacy

[5] https://www.cnet.com/how-to/google-collects-a-frightening-amount-of-data-about-you-you-can-find-and-delete-it-now/
[6] https://www.ericsson.com/en/blog/2020/4/brain-controlled-technology-in-ten-years
[7] https://www.311institute.com/the-worlds-first-telepathic-interface-is-getting-better/
[8] Autonomy of Individuals and Organizations: Towards a Strategy Research Agenda David M. Brock* Department of Business Administration, Ben-Gurion University of the Negev, Israel and Department of International Business, University of Auckland, New Zealand, International Journal of Business and Economics, 2003, Vol. 2, No. 1, 57–73
[9] Blockchain Thinking: The Brain as a DAC (Decentralized Autonomous Organization) Melanie Swan, Contemporary Philosophy MA Candidate, Kingston University London, 2015.
[10] https://www.dezentrum.ch/en/blog/digitalization-is-followed-by-decentralization/
[11] https://neurosciencenews.com/brain-internet-connection-7489/
[12] https://www.openwater.cc/technology
[13] http://www.iflscience.com/brain/brain-brain-interfaces-science-telepathy/
[14] Conscious Brain-to-Brain Communication in Humans Using Non-Invasive Technologies, Charles Grau et al. 2014, http://journals.plos.org/plosone/article?id=10.1371/journal.pone.0105225
[15] http://www.idph.com.br/conteudos/ebooks/BraveNewWorld.pd
[16] https://www.nytimes.com/2019/11/24/opinion/world-wide-web.html
[17] https://webfoundation.org/2020/03/web-birthday-31/
[18] https://contractfortheweb.org/
[19] https://thenextweb.com/growth-quarters/2020/01/30/digital-trends-2020-every-single-stat-you-need-to-know-about-the-internet/
[20] https://journals.sagepub.com/doi/full/10.1177/2056305119863146
[21] https://repository.law.umich.edu/cgi/viewcontent.cgi?article=1294&context=mjrl
[22] https://vetri.global/wp-content/themes/vetri/documents/VETRI_mission.pdf
[23] https://gdpr.eu/
[24] https://www.hhs.gov/hipaa/for-professionals/security/laws-regulations/index.html
[25] https://oag.ca.gov/privacy/ccpa
[26] https://www.weforum.org/communities/gfc-on-data-policy

[27] https://www.nytimes.com/2021/01/10/technology/tim-berners-lee-privacy-internet.html
[28] https://cybersecurity-magazine.com/the-humanized-internet-interview-with-monique-morrow/
[29] https://www.businessinsider.com/whatsapp-privacy-policy-delay-three-months-2021-1?r=US&IR=T
[30] https://www.weforum.org/communities/gfc-on-data-policy
[31] The Ethical Algorithm : The Science of Socially Aware Algorithm Design, Michael Kearns and Aaron Roth, Oxford University Press. 2020.
[32] https://policyreview.info/open-abstracts/self-sovereign-identity
[33] https://ec.europa.eu/futurium/en/system/files/ged/eidas_supported_ssi_may_2019_0.pdf
[34] https://www.europarl.europa.eu/legislative-train/theme-a-europe-fit-for-the-digital-age/file-eid
[35] https://www.w3.org/TR/did-core/
[36] https://www.nytimes.com/2021/01/10/technology/tim-berners-lee-privacy-internet.html
[37] https://www.bag.admin.ch/bag/en/home/krankheiten/ausbrueche-epidemien-pandemien/aktuelle-ausbrueche-epidemien/novel-cov/swiss-covid-app-und-contact-tracing.html]
[38] https://www.npr.org/2019/09/13/760478177/microsoft-president-democracy-is-at-stake-regulate-big-tech
[39] https://www.eweek.com/security/microsoft-ceo-says-privacy-a-basic-human-right-calls-for-global-gdpr
[40] https://www.fastcompany.com/90421349/one-congresswoman-perfectly-summed-up-facebooks-hypocritical-stance-on-privacy
[41] https://www.cnet.com/how-to/google-collects-a-frightening-amount-of-data-about-you-you-can-find-and-delete-it-now/
[42] https://www.ericsson.com/en/blog/2020/4/brain-controlled-technology-in-ten-years
[43] https://www.311institute.com/the-worlds-first-telepathic-interface-is-getting-better/
[44] https://www.nytimes.com/2019/11/24/opinion/world-wide-web.html; https://webfoundation.org/2020/03/web-birthday-31/
[45] https://contractfortheweb.org/
[46] https://edition.cnn.com/2017/02/21/africa/mpesa-10th-anniversary/index.html; https://thenextweb.com/growth-quarters/2020/01/30/digital-trends-2020-every-single-stat-you-need-to-know-about-the-internet/;

https://journals.sagepub.com/doi/full/10.1177/2056305119863146; https://repository.law.umich.edu/cgi/viewcontent.cgi?article=1294&context=mjrl

[47] https://www.weforum.org/communities/gfc-on-data-policy

[48] https://policyreview.info/open-abstracts/self-sovereign-identity

[49] https://www.nytimes.com/2021/01/10/technology/tim-berners-lee-privacy-internet.html

[50] https://ec.europa.eu/futurium/en/system/files/ged/eidas_supported_ssi_may_2019_0.pdf

[51] See also: https://www.europarl.europa.eu/legislative-train/theme-a-europe-fit-for-the-digital-age/file-eid

[52] https://www.w3.org/TR/did-core/

PART II

Societal Dilemmas

Societal dilemmas seek to explore the technology polarities of a Humanized Internet at a human level such as impact on refugees and other individuals in our society. Chapter 3 frames up the narrative and the emergence of organizations that have been applying the use of technologies for credentialing. Chapter 4 focuses on the theme of tokenization of the self and a call for a new social contract. Let us explore how the industry is coalescing on some of these essential problems to create opportunities for those individuals who are under-served.

DOI: 10.1201/9781003339724-5

3

New Organizations

Tey Al Rjula is of Syrian descent, born in Kuwait. In 1990, at the height of the Gulf War, the country's civil registries were destroyed. Lost in the destruction were Tey's birth certificate and other important documents.

In 2011 he came to the Netherlands, where he worked for five years until his contract expired. With Syria gripped by war, returning was not an option. Tey's only chance to stay in his adopted country was to apply for asylum.

But there was a problem. Tey knew who he was. But with the destruction of his records in Kuwait, he could not prove any of it. "There I was in the asylum camp, without a birth certificate, without documents, not being able to identify who I am."

3.1 Introduction

Tey's story is similar to the co-author of this book, Akram. These stories represent a swath of humanity that is displaced due to war, migration, and so on. In fact, it is because of Tey, that I was inspired to matriculate at the University of Nicosia and pursue my M.S. degree in blockchain and digital currency having graduated in June 2019. Tey had shared this program several years ago when we had met via conference call and reunited in Geneva, in January 2019. The program and journey have been most enriching and insightful, blessed with meeting leaders like Tey along the way.

On a personal level, I always commence a presentation with a question, "What would you do if you had minutes to leave? What would you take with you?"

I pose this question often having had the opportunity to be part of an UNHCR visit to the Zaatari Refugee Camp in May 2018.

At Zaatari, I had the opportunity to meet some of refugee families none of whom envisaged their lives to be so disrupted. Many had met their spouses

DOI: 10.1201/9781003339724-6

in the camp and are raising their children in the camp. A world turned upside down by war. These people have dreams to continue their education and professions. The emphasis on education particularly struck me as some people tend to take education for granted. Education, like healthcare, is one of society's benefits if and only if you have an identity that is recognized by the country you live in.

There was no doubt that being able to contribute to society whether as a dentist, doctor, or teacher was very important for the people I had met.

The question we must ask ourselves is whether or not the institutions we have depended upon for societal attribution in the form of passports, university degrees and so on are challenged by these dilemmas. More importantly, what is the impact on democracies as we know it today? We have a world in flux and migrations of people have been constant, further exacerbated by climate change and war.

Human rights are universal, at least from a western constitutional perspective. With the increasing instability of states and economies in recent times, resulting in new waves of migration worldwide and the refugee crisis, In particular, human dignity is under threat in various respects. One very basic aspect is the loss of identity and citizenship of the individual frequently caused by the circumstances of migration. On the other hand, we could lose our papers via natural disasters and other catastrophes beyond our control. Self-sovereign identity means that the citizen has control over his/her own data. Self-sovereign identity is foundational to other services and businesses beyond national borders. It empowers the individual having control over his/her data independently of nationality and place of residence or business. It can help, first of all, to establish citizenship of the person of his/her state of origin. Without this, the regime of statelessness would apply in certain cases, with the consequence that almost no rights are granted to the person, nor diplomatic protection, and so on.

The stakes are higher as these vulnerable people become victims of human trafficking. Think about it, you are stateless, with no papers and exposed to nefarious groups whose livelihood is human slavery. The story of Meena Hassena was sold into sex slavery at the age of 8 by her uncle and captured in a riveting film and featured at ID-2020 in New York, 2016.

Without credentials that attest to who are, when you were born and so on, you have no access to basic services such as education, healthcare, financial inclusion, and so on.

It could be you or me. Self-sovereign identity is the cornerstone for any proposed solution for helping refugees or displaced humans. Simply put, self-sovereign identity is the gate and the first step for social inclusion, financial inclusion, re-settlement, access to services, and preventing human trafficking.

Solving the most pressing issues of our time – 10 1× billion+ people's problems – is indeed an opportunity. In other framing words: what if we eliminate a billion starving moments, a billion moments of frustration, a billion moments of death and suffering etc.?

Can we create a billion moments of happiness, fairness, compassion, empathy, freedom, or opportunity?

Self-sovereign identity is at the core of it all along with the associated personal data that is collected, persistent and portable for the individual.

Another emerging value proposition is "Technology for Social Good." Some examples include AI for good, blockchain for social good, and so on. This theme emphasizes stakeholder vs. shareholder value and is described in the book, *Conscious Leadership, Elevating Humanity Through Business*. The technology for good component intersects with the issue of societal dilemmas to some extent.

Purpose and social benefit are critical values especially for Generation Z leaders, those persons who are born between 1997 and 2015.

"Gen Z is socially focused. The first thing you have to keep in mind about Generation Z is that many are socially focused. They want to support businesses that improve lives on this planet; one McKinsey survey found that 70% of Gen Z respondents try to purchase from companies they consider ethical."

There is an intersection between using technology to develop solutions for the under-served like stateless people, refugees, migrant workers and developing a for purpose socially responsible framework that can be adopted globally.

Let us explore these concepts and organizations in the next section of this chapter.

3.2 Emerging Organizations and Programs

To provide further context to this discussion, I had published a blog in UNHCR to describe the potential of technology, e.g., blockchain, to address some of these examples. The following is an excerpt from the blog itself:

"Identity is a complex and not fully defined concept that nonetheless today plays a central role in the debate within the social and political sciences. Name, age, gender, ethnicity, citizenship, and other social categories all contribute to the definition and uniqueness of the individual, who becomes socially 'recognizable' by virtue of proving those qualities. Armed with his or her unique identity, the individual can then access school, contract marriage, gain employment, own land, receive healthcare and fully participate in the institutions of a national or international community. The increased instability of states and economies underlying the new waves of migration worldwide poses a historic threat to human dignity in many different ways. One very intricate facet of this phenomenon is the weakening – and at times even the loss – of personal identity arising from the fading bonds of citizenship that are frequently the result of the circumstances of migration. Among human groups on the move, refugees are particularly affected by this issue. They often find it enormously challenging to prove fundamental aspects of their identity. The damage and loss of documentation validating their national identity, personal data and information, professional diplomas, and so on create insurmountable barriers for conducting a dignified existence, accessing services and rebuilding their lives. For many, proof of their identity may still be stored in physical or digital archives within the institutions of their failing state. Alternatively, this proof may have forever vanished because those institutions have ceased to exist. The paradox of not owning the proof of your own existence is striking, and it is laden with consequences for refugees. UNHCR registers millions of refugees on behalf of governments around the world, maintains those records in a proprietary database and issues refugee ID cards. This is an invaluable service and a lifeline for refugees."

This blog generated quite a few insightful comments namely around the fear of surveillance and certainly the requirement to avoid posting Personal Identifiable Information or PII to a public ledger which could be the source of abuse by governments.

UNHCR continues its innovation via its technology hub program by engaging refugees to be part of the solution as an inclusive program.

Continuing under the umbrella of the United Nations, *The United Nations* and the correlation to the *Sustainable Development Goals (SDGs)* is perhaps the area to focus on. At a macro level, UN SDG Goal 16.9 has identified the goal to provide a legal identity to all including birth registration by 2030.

"Sustainable Development Goal Target 16.9 ("legal identity for all, including birth registration, by 2030") is key to advance the 2030 Agenda commitment to leave no one behind, and equally relevant is SDG 17.19 – support to statistical capacity-building in developing countries, monitored by the indicator "proportion of countries that have achieved 100 percent birth registration and 80 per cent death registration."

While this goal is visionary, the implications of realizing it, beg several questions with trust being foundational. If we bind the problem area to the portability of digital identities to attestation of digital credentialing, what can be done in developing a trust framework for citizens globally and for policymakers? In fact, one could argue that there needs to be more capacity building on these topics globally. Remember, it could be you.

Yes, it is true that UNHCR and the United Nations are not new and certainly not emerging organizations. Perhaps one could assert that linking the topics of identity to disenfranchised populations could accelerate solutions in this space of digital identity and portability of digital lockboxes.

The assumption ceteris paribus is that the citizen should be the center of this trust universe. This is a noble assertion but do we as citizens believe this to be the case? An aspect of the social dilemma is specific to the idea that a pervasive Electronic Identification or eID will be sufficient. Whilst this topic will be explored in Chapter 12, it should be no surprise that Swiss citizens voted against eID referendum on March 7, 2021, because of the impact on citizen data privacy, shared accountability between the federal government and private firms providing the eID solution was simply not understood by the population. It was not enough to assert that "your participation is voluntary," the possible misuse was what citizens who voted no simply felt. There will be more analysis on the March 7, 2021 eID Swiss referendum, the point here is that the general population did not see the benefit.

There are emerging organizations in the digital identity space. Do note that we distinguish between traditional Identity Access Management solutions that focus on corporate requirements for authentication mechanisms for example and Digital Identity, more specifically self-sovereign identity itself. The latter has been emerging these past several years and concentrates on selective disclosure, but more importantly digital portability of credentialed documents.

There are few organizations that we call out here though certainly this list is non-exhaustive.

One World Identity or OWI has developed a honeycomb of verticals and providers.

"OWI was originally founded on the concept that digital identity was broken, with too many data silos. Over the past five years, the organization has launched about 25 products, and completed close to $4 billion in client transaction services, according to the presentation.

In its first investor forum of the year, 'Riding the Digital Wave to Identity Strategy,' OWI examined the 2021 Identity Landscape from a strategy perspective, looked at investment trends in the space, and spotlighted decentralized identity startup Gataca."

OWI has approached a fragmented digital identity to map use cases with actual solutions. The common theme being "decentralized identity." I was recognized as one of the top 100 influencers in Identity by OWI in 2019.

Another organization is the *Open Identity Exchange* has been developing technology and solutions that focus on some of these examples.

The Internet Identity Workshop or IIW: Co-founded by Kaliya Young and has been labeled as the "Woodstock for Identity." Kaliya is known as "Identity Woman" and has been working on self-sovereign identity for several years now. The Internet Identity Workshop is use case driven with emphasis on solutions and outcomes. As an observation, however, there could be further impact if the workshop could round-robin globally and/or allow for virtual participation. The "who's who" in the identity world participates in IIW.

Self-sovereign identity or SSI: There have been organizations working on SSI, e.g., sovrin.org or the Linux Foundation, with its Feb 2021 announcement of "DizmeID Foundation."

"The DizmeID Foundation and technical project will define and allow for the implementation of Dizme features on top of Sovrin public identity utility. The Dizme ecosystem is expected to include various technological components leveraging Hyperledger stack and adding a monetization layer based on Algorand blockchain protocol, which will enable the exchange of verifiable credentials and the development of new vertical applications. The identity credentials are managed with three levels of assurance: low, self-declared information; medium, automatic checks; and substantial, trusted identification. These levels of assurance would enable industry to have safer, innovative and cost-effective onboarding processes."

Alex Preukschat and Drummond Reed book on self-sovereign identity is a must-read reference for SSI. The book essentially articulates the need for an identity layer in the internet and the use of blockchain and cryptographic primitives to bolster identity decentralization and control by the individual.

"What is our identity? It is everything we do in our daily life, but with the identity data records siloed in banks, big corporations, and governments. Currently, we rely on these third-party records to prove who we are and what we own – but we do not control them.

Self-sovereign identity changes everything.

Imagine selling a house, applying for a driver's license, or checking an applicant's credentials without evaluating a portfolio of papers gathered from various centralized institutions. Powered by secure blockchain registries and cryptography, self-sovereign identity is empowering people across the globe to take back control of their own data using digital credentials controlled by the users themselves. It eliminates the need for organizations to manage customer data directly, reducing costs, and massively increasing security. SSI is creating new, richer, and more trusted relationships online, and promises to be just as important as the emergence of the Internet."

The essential characteristics of SSI are that users control their profiles and what information services can access. Once users have their private keys, they require somewhere to store them without ceding control to a central provider. With self-sovereign identity, you provide only information for specific interactions e.g., whether someone has more than 100,000 USD or is over 21 years old without actually storing the information on the blockchain.

Kaliya Young stated in 2005: "Now I have a way that I can create an ID. I can put it somewhere, and they can't take it away from me. SSI is the promise for the type of societal dilemmas in this narrative of digital identity portability, control and selective disclosure.

3.2.1 Decentralized identity foundation

The decentralized identity foundation (DIF) is engineering-driven with an emphasis on developing technology enablers to decentralized identity. Examples include the creation of protocols and decentralized identifiers (DIDs) for application of DID resource discovery and the allocation, storage of DIDs over a secure DID communication channel. The authentication of DIDs and use of credentialing are in the scope of the DIF activities. DIF is use case driven which leads to practical functional specifications that are required for implementation.

For example:

"The key piece of the decentralized identity equation is how people, organizations, and devices can be identified and located without centralized systems of identifiers (e.g., email addresses). DIF members are actively working on protocols and implementations that enable creation, resolution, and discovery of decentralized identifiers and names across decentralized systems, like blockchains and distributed ledgers.

Creating trust between DIDs and gaining access to products, services, and systems with DIDs requires the acquisition, generation, and inspection of credentials (DID-signed data objects).

The DID Credential Manifest is a format that aims to normalize the process of credential acquisition, wherein the issuer is able to describe the requirements the subject or participant in the credential generation process must meet for the issuer to generate the desired credential."

There are special interest groups, for example, in the financial and healthcare sectors.

The Humanized Internet Organization is also a member of DIF. The Humanized Internet Organization's interest is both in understanding the technology and in contributing to some of the use case examples with an emphasis on credentialing.

The core work related to credentials and decentralized identity is found in the W3C Credentials Working Group.

3.2.2 W3C Credentials Working Group

The W3C Credentials Working Group has been active in credential development that concentrates on verifiable claims since 2014.

> "The mission of the W3C Credentials Community Group is to explore the creation, storage, presentation, verification, and user control of credentials. We focus on a verifiable credential (a set of claims) created by an issuer about a subject – a person, group, or thing–and seek solutions inclusive of approaches such as self-sovereign identity; presentation of proofs by the bearer; data minimization; and centralized, federated, and decentralized registry and identity systems. Our tasks include drafting and incubating Internet specifications for further standardization and prototyping and testing reference implementations."

As one may already observe, there is an intersection between the W3C Credential Working Group, DIF, emerging work in SSI and the overall identity space. I would argue that the W3C has been pivotal in establishing frameworks that can be applied to the credentialing space. Think about it, if you have no proof of education, work experience or even more egregious, are stateless, could we not use these technologies to verifiably provide these proofs? The response is "yes," and even more compelling is for these proofs to be portable. Indeed, when a verifying organization no longer exists and/or is replaced by another entity, these claims should persist. The static form of proof of x or proof of y is no longer sustainable in the 21st century hence why a novel application of credentialing is required.

Sir Tim Berners-Lee, Solid Project, the Solid project takes credentialing and portable identification a step further. Solid advocates for true data ownership.

We note the relationship between data governance and the notion of data ownership, to identity, contextually, data about you.

"Users should have the freedom to choose where their data resides and who is allowed to access it. By decoupling content from the application itself, users are now able to do so."

For someone who is displaced or disenfranchised, the topic may be rather ethereal. In fact, if we connect the dots further, this area of data ownership is a very difficult problem to solve and indeed pertains to all of us. Data about me linked to my identity that is coupled with persistent self-sovereign identity to verifiable claims form a loop where the human is the center. I am in the center.

Remember the question that I posed earlier, if you had minutes to leave due to a disaster, what would you take with you?

You can even get your own data pod from Solid.

The operational arm of Solid is Inrupt which is a startup tasked to develop the solutions for Solid and is based on open source.

The overall Solid-Inrupt amplification is how you wish to manage data about you to a point that cannot be lost when addressing the topic of societal dilemmas.

The Biometrics Institute challenges us to think about the responsible use of biometrics as this technology becomes rapidly adopted by consumers and enterprises as a source of identity verification.

"The Biometrics Institute is an independent and international membership organization for biometrics. It was founded in July 2001 in Australia responding to an industrial need for an impartial forum for sharing knowledge and information about biometrics. It expanded its outreach to New Zealand in 2004, to Singapore in 2011 and opened a subsidiary in London, UK, in July 2011 to create a global biometrics network."

Can we imagine the abuses of facial recognition and any biometric use of data to our identity? There is not a lack of these abuses and certainly what may be interpreted as an abuse can be defined as a state surveillance or perhaps a new definition of societal harmonization. Within the context of societal dilemmas, how we define and agree to biometric use of data is daunting and can quickly lead to a very dystopian narrative.

Biometric based Digital Identity Initiatives have been emerging in Africa and India, ID4Africa and the Aadhaar project for the past several years. These national initiatives fall under the UN Sustainable Goal, 16.9, for legal identity for all by 2030. These projects have not been without challenges but we do note that national identity initiatives are commencing with biometric mechanisms. What we need to postulate is how far can a national biometric database go, e.g., your DNA and understand the potential for abuse.

What about behavioral biometrics for eye tracking, keystroke pattern development, and so on to develop a profile about you or I? This is a contentious topic for sure and is an area that requires us all to track further.

3.3 A Word About Data Policy

Assessing patterns under societal dilemmas data policy is part of this discussion. Remember that data and identity are linked. Data about you is contextual and identity is complex. At a fundamental level, I have multiple identities, a personal identity, a professional identity, an ethnic identity, a sexual identity, a cultural identity, and so on.

However there is no common data policy in this discussion. In fact, the World Economic Forum (WEF) has established a Data Policy Council 2020–2021 to define common attributes that may correlate to a data policy.

The WEF Data Policy Council's focus is:

"To unlock the value of data, organizations, whether in the public, private or third sector, must examine the ways in which they store, integrate and process data. Furthermore, there is untapped potential in data that, if harnessed, can have implications for helping to solve societies' biggest problems or creating new economic value.

Data architecture and systems design can help solve this issue, going beyond simple core principles into the technical execution of logical processes to ensure functionality and innovation.

The Global Future Council on Data Policy will examine data architecture models that promulgate an appropriate balance of creativity, responsible use and efficiency."

I am a member of the WEF Data Policy Council 2020–2021 and can attest to the difficulty in defining a common data policy framework. We agree that the topic is contextual in nature. We have been gravitating to understanding what the responsible use of data can look like. The intersection to privacy, cybersecurity, artificial intelligence, machine learning, and deep fakes are among the numerous topics the council members have been discussing. Candidly, we are at the tip of the iceberg on data policy that can be understood and embraced globally.

3.4 Technology for Good: Closing in on These Social Dilemmas

The social dilemma between technology for good or for purpose often is deluded into corporate social responsibility (CSR) organizations within large enterprises. There is no doubt that positive work is being done by these groups. Under the context of digital identity and its impact on individuals, McKinsey researched the correlation of digital identity [perceived for good] to value creation for corporations and countries. I often categorize this type of value creation as "Technology for Good."

Technology for Good needs to be embedded in an organization's DNA in order for it to be concrete and truly authentic. There are some examples specific to impacts of technology, our climate, food, water, and disenfranchised people, e.g., the latter regarding digital identity and credentialing as previously articulated.

However, with every opportunity there is a warning on intentional use and abuse.

"Digital ID, much like other technological innovations such as nuclear energy and even the ubiquitous GPS, can be used to create value or inflict harm. Without proper controls, digital ID system administrators with nefarious aims, whether they work for private-sector firms or governments, would gain access to and control over data. History provides ugly examples of misuse of traditional identification programs, including tracking or persecuting ethnic and religious groups. Digital ID, if improperly designed, could be used in yet more targeted ways against the interests of individuals or groups by the government or the private sector. Potential motivations could include financial profit from the collection and storage of personal data, political manipulation of an electorate, and social control of particular groups through surveillance and restriction of access to uses such as payments, travel, and social media."

I could imagine that we will witness widespread development of investment tools in this space already defined as Environment Social Governments or ESG.

Technology for Good really translates into an evaluation of how technology can impact society positively. Articulating the intentional use of a technology and perhaps its potential for abuse may be starting points in this discussion. There is an opportunity to develop a Technology for Good framework with metrics that prove a benefit for both enterprises and society. I do believe that self-sovereign identity could be part of this Technology for Good framework.

Think about the narrative thus far and the organizations that are addressing some aspects of this space.

Sounds like an opportunity, I hope you agree.

3.5 Summary

We have discussed emerging organizations specific to identity, data policy and Technology for Good in Chapter 3. If you have detected some mode of excitement here, it is because we are at the tip of developing what is possible. The call for action is for you, to engage in these organizations and in these discussions. Become an active player and creator of solutions that can and should benefit our society. Do think about self-sovereign identity as a potential starting point. In Part III, we will discuss the Humanized Internet enablers and double-click on ethics, standardization and regulatory challenges.

References

[1] https://win.systems/stories/toufic-tey-a-rjula/
[2] https://www.henleyglobal.com/newsroom/latest-news/inside-the-refugee-camp-where-technology-improves-lives
[3] http://www.meenathemovie.com/
[4] Conscious Leadership, Elevating Humanity Through Business, J.Mackey et al., Penguin, Whole Foods Market Services, Inc, 2020
[5] https://www.unhcr.org/blogs/promise-hype-provides-blockchain-safe-identity/
[6] https://www.unhcr.org/innovation/
[7] https://sustainabledevelopment.un.org/?menu=1300
[8] https://unstats.un.org/legal-identity-agenda/
[9] https://www.bk.admin.ch/ch/d/pore/va/20210307/can639.html#]
[10] https://oneworldidentity.com/2021-digital-identity-landscape/
[11] https://www.biometricupdate.com/202102/new-digital-id-entrants-and-growth-opportunities-appear-on-owis-2021-identity-landscape
[12] https://oneworldidentity.com/owi-announces-2019-list-of-top-100-influencers-identity/
[13] https://openidentityexchange.org/members/anon/new.html?destination=%2Findex.html
[14] https://internetidentityworkshop.com/
[15] https://www.linuxfoundation.org/en/press-release/linux-foundation-announces-dizmeid-foundation-to-develop-and-enable-a-self-sovereign-identity-credential-network/
[16] https://www.manning.com/books/self-sovereign-identity
[17] https://identity.foundation/interop/
[18] https://www.w3.org/community/credentials/charter/
[19] https://www.w3.org/community/credentials/
[20] https://solidproject.org/users/get-a-pod
[21] https://www.inrupt.com/solid
[22] https://solid.mit.edu/
[23] https://www.biometricupdate.com/202010/biometrics-institute-declares-three-laws-for-the-responsible-use-of-biometrics
[24] https://www.npr.org/2021/01/05/953515627/facial-recognition-and-beyond-journalist-ventures-inside-chinas-surveillance-sta
[25] https://www.biometricupdate.com/id-for-all
[26] https://www.biometricupdate.com/biometric-news/behavioral-biometric-news-articles]

[27] https://www.weforum.org/communities/gfc-on-data-policy
[28] https://www.mckinsey.com/business-functions/mckinsey-digital/our-insights/digital-identification-a-key-to-inclusive-growth#
[29] https://www.forbes.com/advisor/investing/esg-investing/

4

Societal Vulnerabilities and Implications Toward Tokenization of the Self

Monique J. Morrow[1] and Mehran Zarrebini[2]

[1] Vetri-Global, Klosbachstrasse 128, CH-8032 Zürich, Switzerland
[2] PFE International Inc., 18 Grayston House, 21 Astell Road, London, United Kingdom

Email: monique.morrow223@gmail.com; mehran@zarrebini.com

> "*In the absence of synthetic declarations that secure the road to a human future, the intolerability of glass life turns us toward a societal arms race of counter-declarations in which we search for and embrace increasingly complex ways to hide in our own lives, seeking respite from lawless machines and their masters*"
> [The Age of Capitalism, Shoshana Zuboff, Hachette Book Group, 2019, p. 489]

We are living in a world where the very systems upon which trust is based are being challenged by new and exciting paradigm shifts. Centralization, whether in the form of governments, financial institutions, enterprises, and organizations, is simply being challenged because of the lack of trust associated with data governance often experienced in the form of data breaches or simply a monetization of our data without our permission and/or incentives to participate in this emerging decentralization of structures. We see this trust deficit challenging the very institutions we have depended on including but not limited to financial institutions, private enterprises, or government bodies. A new "social contract" is required as we continuously evolve into more decentralized and self-governing (or semi-self-governing) entities. We will see more development in Digital Sovereignty with the caveat that a governance model will need to be defined.

DOI: 10.1201/9781003339724-7

4.1 Introduction

There are several key areas that will be driving decentralization and, to some extent, the creation of autonomous organizations. For example, the exponential growth of Fifth Generation Technology (5G) and the Internet of Things (IoT) connected devices that possess their own identities and data. They will communicate with each other without human intervention. The IoT industry is ripe for the next major wave of disruption through decentralized applications.

Examples of decentralization include supply chains that are inherently complex, lack transparency, and result in a duplication of tasks among stakeholders impacting efficiency and cost. Other examples include energy companies that are also using digital assets in new and revolutionary ways and may usher in a new era of energy management worldwide such as the use of energy and an exchange of energy value among communities [1]. These assets could later be used to influence how such projects are operated, allowing stakeholders to participate in key decisions that affect their communities [2]. Healthcare and the tokenization of health data, allowing individuals to derive value from their own information, will become increasingly important. This will perhaps challenge the very healthcare insurance entities that we know today. We are moving beyond a tokenization economy but rather witnessing the democratization of data at a societal level.

In this chapter, we explore the distinction between tokenization and encryption and the shift toward decentralization, specifically in the context of human data. We then explore the evolution of the social contract in a digital world where a digital citizen is the source of the social contract. Self-tokenization and data democratization are considered together with the emerging solutions that seek to mitigate data leakage and the protection of human data. Tokenization will impact the way organizations are governed in various ways, and regulatory models of corporate governance will need to be re-examined in light of the profound disruption to business caused by tokenization. We conclude the chapter with a discussion on tokenization and waste management with a specific emphasis on supply chains and autonomous/electrical vehicles.

4.2 Tokenization and Encryption

Digital transformation, "high tech" including AI and blockchain, has provided a multitude of opportunities for organizations to innovate by creating a leap in buyer value and delivering increased efficiencies across a wide cross-section of buyers. Individuals and businesses have therefore benefited tremendously from this transformation but have also become susceptible to security threats and cyber criminality. In a study conducted by the Verizon Business RISK Team, more than 280 million payment card records were breached in 2008 [3]. The global financial market collapse in 2008 created not only economic headwinds of unprecedented proportions but also some of the largest data breaches in history. In fact, the study confirmed that if reasonable security controls had been in place at the time of the incident, the breaches could have been avoided in their entirety.

Compliance with security standards is a resource-intensive challenge to all businesses irrespective of their size. Despite the enormous efforts and vast expenditure of businesses to secure their data, hundreds of millions of records have been breached and continue to do so. For instance in 2008, the Verizon Business RISK team highlighted that retail, financial services, and food and beverage accounted for three-quarters of the 2008 breaches. The majority of the records were compromised from servers and applications [3]. In 66% of the cases, the breach involved data that the organization didn't know existed on their servers. Data breaches have continued to plague numerous industries since 2008. In 2018, an unprecedented number of breaches occurred varying in severity. Hacking groups injected malicious code into poorly secured web pages of British Airways which exposed the data of 3,80,000 customers, and in India, an ID database managed by the Indian Government was compromised exposing the identity details and private information of 1.1 billion individuals [4].

Two technologies utilized in the protection of data are encryption and tokenization. Tokenization and encryption are often mentioned together and their definition is used interchangeably. While both are utilized as data obfuscation technologies, there are clear distinctions between the two. Both technologies can be utilized to secure data under varying circumstances, and in some instances such as end-to-end payment facilitation, both encryption and tokenization are used together.

Encryption is the process of using an algorithm to transform plain text information into a non-readable form called ciphertext. An algorithm and an encryption key are required to decrypt the information and return it to its original plain text format. In symmetric key encryption, the same key is used to both encrypt and decrypt the information. In asymmetric key encryption, two different keys are used for encryption and decryption. Therefore, the private key can decrypt messages sent to an individual but cannot decrypt what is sent to another part as that is encrypted with another key pair. Today, secure sockets layer (SSL) encryption is commonly used to protect the information as it is transmitted on the internet. Using built-in encryption capabilities of operating systems or third-party encryption tools, millions of people encrypt data on their computers to protect against the accidental loss of sensitive data in the event their computer is stolen. Data encryption can be expensive and a resource-intensive proposition.

Tokenization, on the other hand, is the process of converting a piece of data into a random string of characters known as a token. Tokenization protects sensitive data by substituting non-sensitive data. The token serves merely as a reference to the original data but cannot be utilized to determine those values. A credit card number, for example, is replaced within the merchant's storage environment by a token value generated in such a way that it cannot be linked back to the original data element [5]. The token value can, therefore, be used in numerous applications, some of which are considered in this chapter, as a substitute for real data. The advantage of tokens is that there is no mathematical relationship to the real data that they represent. The real data values cannot be obtained through reversal, and hence, a breach renders the information invaluable.

Tokens are being increasingly used to secure varying types of sensitive information. In particular, personal identifiable information such as healthcare information, email addresses, and account numbers are such examples. From a security perspective, tokenization significantly reduces risk based on the fact that sensitive data cannot be breached if it is not there in the first place. However, there are a number of other use cases where tokens create value such as the tokenization of traditional financial assets where liquidity creates barriers to entry and tokenizing these assets, for instance, using blockchain to convert rights into a digital token backed by the asset itself can solve this problem. We are likely to see an increase in the number of tokenized real assets including but not limited to real estate where investors are able to own portions of real estate and collectibles such as art [6]. The future will result in an environment of greater personalization and customization. In both instances, the token represents distributed ownership of the underlying asset's value but not the asset itself, democratizing the process of ownership.

A new "tokenized" economy offers tremendous potential for creating a more efficient and inclusive environment in which tangible and intangible assets can be traded through greater liquidity, accessibility, transparency, and faster and cost-effective transactions [7]. Furthermore, tokenization in essence allows you to compartmentalize personal data and manage it across different users simply and effectively. That data can therefore only be accessed by an entity that has the correct token. The information is secure and completely portable since the personal data are controlled by the individual and shared with permission, a stark difference to the data that is being utilized currently.

4.3 The Shift Toward Decentralization

In addressing the shift toward "decentralization", it is important to define nomenclature and context. For example, the question for clarification is: what is meant by "decentralization" in comparison to "distributed" and "peer-to-peer" constructs? Distributed within the context of organizations has been the topic of a body of research specifically around the study of organization computing [8]. The underlying network and IT applications facilitate collaborative communications. Peer-to-peer (P2P) architecture refers to the foundational network of nodes where the clients do not share any of their resources such as storage capacity, computing power, network connection, bandwidth, and content [9]. Within this chapter, decentralization may assume the use of distributed and P2P capabilities. At a societal level, the shift toward decentralization has various sources such as the financial crisis of 2008, the numerous breaches of citizen data in the form of hacks, and the lack of privacy by design [10].

The thesis is that centralization of citizen data is subject to abuse. This centralization can take the form of government and commercial entities where the governance model is questionable due to these abuses. The exhortation that "data are the new oil" only amplifies the value of citizen data and interpretation that such data are indeed a tokenized asset [11]. The Satoshi Nakamoto paper describing Bitcoin as a "P2P Electronic Cash System" is a manifestation of decentralization at a societal level that offers a frictionless exchange of value [12]. This notion of decentralization goes further to include the creation of "Decentralized Autonomous Organizations" (DAOs) where smart contracts create business and organization logic to provide governance in the absence of centralization [13]. A smart contract is a computer program code that is capable of facilitating, executing, and enforcing the negotiation or performance of an agreement using blockchain technology. The first experiments with DAOs and smart contracts clearly show that these solutions function in an arena that is almost entirely unregulated, and attempts to experiment with DAOs have not gone without controversy and have in themselves stimulated research in this space [14]. While technology possesses no agency and therefore cannot prevent miscreant actors from exploiting its intentional use, there is a chasm developing between intentional/unintentional surveillance by central authorities and the balance of what may be best for society [15, 16].

This observation begs the question as to how a chasm may be resolved and by which entities. What kind of social contract is required for the digital world we share? How is trust defined and what is the role of trust toward decentralization? Social contract theory can be best defined as shaping policy consensus between designated authorities, for example, government entities and citizens. There is an implication that values and ethics are shared [17]. In the 21st century, fostered by the use of social media and the internet, social contracts have been evolving into identity politics and the notion of multiple identities, especially when referring to IoT and to the Internet of Everything and Everyone. Dignity and the struggle for recognition are provoking the requirement for a new social contract that can function in a digital world [18]. Given the pattern of decentralization or a hybrid therein, a novel and relevant social contract must put the citizen in the center of the digital universe as an active player. A citizen can be defined as an individual who selectively and knowingly asserts his/her rights in a digitally driven world. Therefore, a citizen does not necessarily mean an individual who belongs to a particular nation state nor is world citizen implied here. A digital citizen whose data are valuable must be the source of the social contract between him/her and entities that he/she "trusts" preferably in decentralized modalities. This is a new assertion given that 21st century social contracts have been evolving to include Universal Basic Income to the right of guaranteed employment under Industry 4.0. There is still a dependency on government entities where social contracts involve the instrument of taxation that is required to provide a "collective welfare" for nation-state citizens [19]. Who creates the rules and how such rules are implemented may be the function of a societal smart contract that is developed by the digital citizen.

Blockchain has been evolving over the past 10 years to include adoption by enterprises and, in some cases, Governments. However, the missing link has been a digital citizen driven social and smart contract. There is an assumption that the digital citizen possesses a priori knowledge in both the development and use of these emerging mechanisms that foster the creation of a new social contract. This assertion may be true, and therefore, further exploration on the type of training and education and by which entities is recommended.

4.4 Trust in a Centralized World and Emerging Solutions for the Protection of Human Data

There is a body of research that seeks to define "trust" which is a complex sociological, psychological concept and the basis for human interactions [20].

As the old adage goes, “Trust takes years to build, seconds to break, and forever to repair,” [source unknown]. Given the data breaches and misuse of individual/citizen data described in Section 4.3, trust in a centralized world is problematic as the governance model assumes self-regulation by the central parties. This fact is described in the Netflix documentary entitled, “The Great Hack,” where privacy and abuse of individual/citizen data is the tip of an iceberg of a calculated strategy to both take data and control away from the individual [21]. An observation has been that the “outrage” from this data abuse has not come from individual/digital citizens but rather from nation-state governments and privacy, human rights organizations [22]. This fact begs the question as to whether or not the individual is immune to such data abuses that involve his/her data. Perhaps, the response to this hypothetical question may be the degree of perceived or actual abuse, for example, personal bank account and funds stolen versus identity fraud versus data leakage with no “direct harm” as a result of an organizational breach. If the degree of perceived or actual abuse is greater than “no direct harm,” then the question is whether or not the egregious act has resulted in the individual behavioral change from passive to active digital citizen. There is an abundance of research on human hacking, but in terms of empirical and behavioral analysis that confirms a change as a result of such abuse, the research is lacking at least at a macro level [23]. This particular question may be difficult to ascertain because individuals (collectively) may be disinclined to share the fact that they have been victims. Therefore, the postulation is that behavioral change is possible when correlated with the severity of the abuse.

However, trust can manifest itself in a Thomas Hobbes like Leviathan where a central authority like a government entity operates in the background to ensure that society behaves according to the rules that have been developed by the central authority [24]. As social psychologist Thomas Tyler observed, rules will be followed if the citizens perceive procedural fairness [25]. At the opposite end of this trust spectrum is peer-to-peer (P2P) trust, which is based largely on relationships and shared ethical norms. P2P trust may be interpersonal as experienced among families, friends, or via a shared set of operating principles, for example, the internet. Furthermore, the modality of the engineers who develop protocols and architectures to evolve the internet (The Internet Engineering Task Force or IETF) is “rough consensus and running code” and “we reject kings, presidents, and voting” [26].

Additionally, reputation graphs can reflect P2P trust relationships [27]. Blockchain has often been referred to as “trustless trust,” an assertion from Reid Hoffman, and evidenced by the operational nature of blockchain, for example, “in proof we trust” [28]. Within the context of this chapter, there is evidence to suggest that the digital citizen is evolving toward a hybrid approach between P2P and Leviathan trust where the result is the removal of intermediary actors and the requirement for transparency in Leviathan trust models [29]. There are emerging solutions that seek to mitigate against data leakage and the protection of human data. With privacy by design as a tenet, in regulation, the European Union General Data Protection Regulation (GDPR) has defined data protection specific to enterprises. Failure to adhere to GDPR compliance will be expensive for enterprises [30]. GDPR further poses the question as to what constitutes actual data, for example, hashing of personal data within blockchain could be problematic. One recommendation has been to avoid hashing Personal Identifiable Information (PII) as a practice [31]. There has been a consideration for the use of Zero Knowledge Proof (ZKP) mechanisms to mitigate against data leakage where range values can be applied in lieu of actual human-personal-digital citizen data. Commercialization of such capabilities is in progress [32].

Additionally, there is continued dialogue between academics and enterprises as to use cases and standardization of ZKP mechanisms [33]. Secure Multiparty Computation (SMPC) is currently being applied for privacy-enforced computation, on/off chain within the blockchain context [34]. Sir Tim Berners-Lee's project, Solid, is taking a step further in re-designing the web where the individual can be in control of his/her data [35]. Professor Bryan Ford at EPFL in Switzerland has established a Decentralized and Distributed Systems Lab (DEDIS) with "no single trusted party" [36]. David Chaum has been in the privacy space for decades and is now implementing Elixxir [37].

This list of emerging solutions is in no way a comprehensive one but confirms rather the trends toward digital-citizen control of his/her data and the implementation of human data privacy overall. There is room for additional research as to ensure selective disclosure and control for the digital citizen.

4.5 Governance and Regulation in a Tokenized World

Tokenization will impact the way organizations are governed in various ways. Basic notions inherent with corporate governance will be transformed such as accountability, responsibility, transparency, and trust [38]. There is an opposing view that there is no urgent need to put digital technologies on the regulatory agenda as the practice of governance will not be disrupted by technologies such as distributed ledgers, blockchain, and IoT. Clearly, the products and services of these organizations will be affected but traditional regulatory models are robust enough to deal with such disruption [39].

There is an emerging view that extant regulatory models of corporate governance will struggle and need to be re-examined in light of the profound disruption to business caused by tokenization [39]. Tokens can be created to develop a link between the token and economic or non-economic ecosystem such as personalized data and can also be created to record or transfer these data via computer code on the distributed ledger. The general topic of tokenization is less than 3 years old and that of self-tokenization, a relatively new topic. The digitalization of all kinds of both tangible and intangible assets and of the creation of new types of rights raises substantial governance issues. It has been shown that with the tokenization of tangible assets such as securities, the new corporate stakeholders or token holders could potentially affect the balance of power within organizations [40]. Furthermore, a new form of trust, digital trust, based on mathematical algorithms and machines, is being developed as the reliance on intermediaries and third parties diminishes. The rapid pace at which the development of the distributed ledger and blockchain is taking shape reinforces the dependency on digital trust. Therefore, digital trust has an impact on how we interact with other entities and society in general. People will increasingly trust decentralized forms of governance in the future [41]. Libra aims to create a reliable digital currency and infrastructure that can deliver on the promise of the "internet of money". To make the mission of Libra, a stable cryptocurrency governed by an independent association tasked with evolving the ecosystem, a reality, a governing entity, the Libra Association [42] has been designed to facilitate the operation of the Libra blockchain. The association will develop and adopt a comprehensive charter and a set of bylaws for the association on the basis of the currently proposed governance structure. The association will ultimately develop a path toward permissionless governance and consensus on the Libra network.

Blockchain technology has the potential to solve principal-agent issues that flourish in the area of regulation and governance. Individuals would be able to verify and regulate access to, and use of, personal data. Legislation would have to adapt in order to clearly define data ownership and the use of non-consensual data. In the area of personal data, there is a tradeoff between the claim that private data ownership would enhance privacy protection and sovereignty over one's own data and the public's need for data access in instances such as enhancing researchers' access to data for clinical care, public health, and research [43].

Increased transparency has the potential to protect individuals. Personal data storage on the blockchain will allow people to own and maintain personal data, granting and rescinding access, and enabling it to be used, shared, or deleted as necessary. Each user has complete transparency over what data are being collected and how they are processed [44]. Disintermediation would simplify the process of authorizing access to personal data, which would streamline processes and increase efficiency. With increased mobile application usage, a set of permissions are often granted upon download or sign-up. While opting-out is generally the only way to relinquish continued access to data, the improvement of existing permission dialog in mobile applications, access control policies will be stored on the blockchain where the user has rights to modify them or revoke access to previously collected data.

4.6 Tokenization in Waste Management, Electrical and Autonomous Vehicles

A circular economy is an industrial system that is restorative or regenerative by intention and design [45]. The "end-of-life" concept is replaced with the concept of restoration and strives for waste elimination through the innovative design of materials, products, systems, and introduction of novel business models. Through numerous technological advances, recycling has become increasingly more viable through the deployment of technology such as Radio Frequency Identification (RFID) and IoT. The deployment of these technologies has created greater efficiency with respect to logistics, knowledge sharing, and tracking of materials [46].

More recently, various circularity initiatives have begun leveraging from blockchain technology for verification purposes. RecycleToCoin [47] looks to increase participation through recycling waste in the United Kingdom. Users of the RecycleToCoin application are incentivized to take their recyclable plastic and aluminum cans to participating stores. They are then given tokens in a secure digital wallet based on the quantity of recyclable material deposited which can then be turned into digital currencies (Bitcoin or Ethereum) or gift cards via GiftPay [48] or donated to PlasticBank [49]. PlasticBank has created recycling systems in many global locations providing fair value for waste collectors and supporting digital transactions through the employment of a permissioned blockchain. In order to achieve scalability and incorporate other technologies such as analytics and visual recognition, PlasticBank has employed blockchain on IBM LinuxONE [50].

Food waste throughout the world is an increasing problem due to its impact on the environment, natural resource scarcity and significant contribution to global warming and climate change. One-third of all food produced is lost or wasted. The United Nations Sustainable Development Goals (SDGs) aim to include a target to halve per capita global food waste at the retail and consumer level and reduce food losses along production and supply chains, including post-harvest losses by 2030 [51]. In Europe, food waste is an increasing concern. The production, distribution, and storage of food use natural resources, and the discarding of food, especially which is still edible, increases these impacts. One company in the United States is aiming to tackle food waste in order to solve food insecurity and also resolve an estimated USD 40 billion in tax credits for businesses that remain unclaimed annually. Goodr [52] is utilizing blockchain technology to track an organizations' surplus food waste from collection to donation, enabling tax deductions for organizations fully compliant with the Internal Revenue Service (IRS) in the United States, reducing greenhouse gas emissions and creating social good by moving edible surplus to communities in need.

Blockchain technology can also be used in other waste management industries and plays a pivotal role in ensuring that recyclables do not end up in landfills. Digital tracking data allow for deeper analyses of supply chains. Utilizing this technology with IoT devices and RFID, greater efficiencies can be gained and rogue activity marginalized. Such activity has been prevalent in the tire recycling industry across the globe. Each year, 1.6 billion new tires are generated and around 1 billion of waste tires are generated [53]. The collection and recycling of tires and prevention from landfill, sensitive habitats and abandoned areas remain a challenge for the industry globally. The tire supply chain consists of the production of raw materials, production of tires, distribution of tires, use of tires, and then the collection, sorting, and recycling of end-of-life tires (ELTs). ELTs are processed into rubber granulate and energy recovery.

There are therefore a number of stakeholders involved in the industry which include raw material producers, tire manufacturers, buyers and sellers, waste collectors, recyclers, logistics companies, and the Government who oversee the recycling industry from a regulatory and governance perspective. In the supply chain, the raw materials constitute the basic unit out of which a tire is produced. This is the basic material from which a tire is created. The tire manufacturers could therefore tokenize the raw materials. The token would represent a digital twin that is a depiction of the raw materials in the form of a token utilised to trace the raw materials throughout the supply chain. This allows stakeholders to follow the newly defined asset throughout the supply chain.

There are a number of ways to digitalize the raw materials in a tire. The raw materials could be deemed a non-fungible asset, an asset in the form of a token which through cryptography would help to verifiably prove ownership and authenticity of the asset. Another option is to have the asset defined as a fungible asset, and the asset can therefore not be interchanged. Therefore, if a specific batch of raw material is required to be traced throughout the supply chain, then this approach would be more feasible. In this instance, the raw materials will be assumed to be non-fungible. The raw material producers therefore generate tokens which are then supplied to the logistic companies. The logistic companies take the raw material and the equivalent amount of tokens representing the amount of raw materials in weight and transport or ship them to the tire manufacturing plant. The manufacturing plant receives both tokens and raw materials and produces tires. The tire producer can therefore continue with the token that represents weight or the token can be converted to represent the actual number of tires produced.

The raw material tokens can therefore be spent in exchange of a tire token that will be issued by the manufacturing plant. Another option would be to create a batch token that represents the weight of raw material utilised to make a batch of tires. This batch of tires, once manufactured, is moved to a warehouse. A seller of the tires has both a batch of tires for sale and the equivalent number of tokens that represent the batch of tires. As tires are sold, a quick response code could be scanned to verify the authenticity of the tires in the supply chain. Additionally, the buyer of the tire could be provided with the token that verifies the authenticity of the tires purchased. As the tires are replaced in the future and ELTs transported to waste tire processors, there is full visibility and traceability along the entire supply chain. The tokens therefore represent the validation of the genuineness of the product itself.

With multiple stakeholder involvement including multiple raw material manufacturers, tire producers, logistics companies, and recycling companies, the privacy of information would be of great concern. The privacy of each stakeholder could be controlled through the private issuance of tokens utilizing ZKPs [32]. Certain information could be kept private in order to protect different stakeholders, and visibility can be granted to entities on a need to know basis. Ownership of blockchain assets can be transferred without revealing confidential transaction details while still ensuring regulatory compliance according to custom-defined business rules. In the event that a Government required oversight of the supply chain, they could be provided with an auditor role utilizing viewing keys in order to gain industry insights into the network. Such insights would include knowledge of productivity, manufacturing outputs, prohibition of resale of waste tires, and output validification from recyclers to determine processing subsidies. The provision of end-to-end visibility into multiple enterprise supply chain networks utilizing blockchain technology enables multiple stakeholders to track and trace end-of-life tires creating trust by recording indisputable transactions on the distributed ledger.

Electric vehicles represent the fastest growing sector of vehicle sales globally, with sales increasing by 64% in 2018 and Tesla's Model 3 Sedans the biggest seller in the second half of 2018. In China, sales of electric vehicles increased by over 5,00,000 units to 1.2 million in 2018, by far the largest contributor to electric vehicle sales [54]. At the end of 2018, the global fleet of light vehicle plug-ins was 5.4 million with an estimated sale of 3.2 million vehicles in 2019 [54]. Power Ledger and Silicon Valley Power [55] have recently completed a blockchain trial that monetizes electric vehicle charging infrastructure, creating the potential for tokenized carbon credit trading. Power Ledger's blockchain-backed platform was trialled to track and manage the Low Carbon Fuel Standard (LCFS) credit generation from solar panels and electric vehicle charging infrastructure, as well as credit trading. The solar and charging data produced from the Tasman infrastructure were recorded as smart contracts on the blockchain platform, which then created a digitized LCFS certificate. This certificate contained a cryptographic hash of the information calculated in real time, which was then sent to Silicon Valley Power's wallet, enabling Silicon Valley Power to monetize their electric vehicle charging infrastructure, thereby creating a secondary market for tokenizing and trading LCFS credits.

The rise of autonomous vehicles is set to be transformational as the market is set to reach USD 42 billion by 2025 [56]. While ethical challenges with respect to split second decision-making remain a major concern, greater understanding of the ethical guidelines artificial intelligence (AI) will follow need to be understood in great depth. When driving, individuals learn from their own mistakes. They rarely learn from others, collectively making the same mistakes over and over again. AI, on the other hand, evolves differently. When an autonomous vehicle makes an error, all of the other autonomous vehicles are able to learn from it. In fact, new autonomous vehicles will inherit the complete skill set of their ancestors and peers; so collectively, these cars can learn faster than people. If these data are recorded on the blockchain, new autonomous vehicles essentially evolve at a rapid pace since the data are available immediately via the distributed ledger.

It is inevitable that in a short period of time, autonomous vehicles will maneuver along roads in conjunction with human-driven cars. Sophisticated AI tools and distributed ledger technology will empower individuals to better learn from the experiences of others. The increased use of autonomous vehicles is ultimately premised on trust. The individuals who buy or use them have to trust the technology and must be comfortable using that technology for their true value to be realized. In order to build this trust and acceptance, autonomous vehicle manufacturers must ensure the technology is safe and secure and creates a leap in value for consumers. The machine world and social world will therefore operate in unison. Humans will therefore emulate the superior learning processes of the machine. The machine hive will become a role model for a new human hive in which we march in peaceful unison toward a world free of mistakes and accidents [15].

4.7 Conclusion

In summarizing this chapter, the authors explored the application of tokenization to the individual and assessed its societal implications in the form of defining a new social contract, articulating the security, privacy in terms of encryption, and identifying further mechanisms as to further secure digital citizen data and assert control. It was shown that at the societal level, blockchain technology is being used in the waste management industry in combination with IoT devices to reduce waste and greenhouse gas emissions and create social good. It was demonstrated that blockchain technology can play a pivotal role in ensuring that recyclables such as tires do not end up in landfill. Digital tracking data allow for deeper analyses of complex supply chains where multiple stakeholder involvement is inherent in the supply chain. The review covered the monetization of electric vehicle charging infrastructure with the advent of global growth in sales of electric vehicles. The application of IoT and blockchain was extended to autonomous vehicles which rely on machine learning and pattern recognition, approaches from the field of artificial intelligence. Blockchain will facilitate the learning approach, requiring an extensive learning strategy and trust where vehicles will be exposed to a large number of complex situations.

Trust architecture models whether Leviathan in nature or P2P will more than likely be hybrid in nature as will the related governance models. There are emerging solutions whether the commercialization of ZKP mechanisms or David Chaum's Elixxir which should be followed. The assertion "we are valuable" implies that the digital citizen must be an active player for selective disclosure and control. There is room for empirical research to confirm the results of this assertion. John Cheny-Lippold's book, *We Are Data, Algorithms and the Making of Our Digital Selves*, [57] is a step in this direction when discussing "data about data". Identity is foundational in this discussion whether referring to an individual, IoT, or connected cars.

In summary, this specific research topic "tokenization of the individual and societal implications" is still nascent.

References

[1] Finance 4.0 (2018). Technical Report. ETH Zurich, Switzerland. Available at: https://www.research-collection.ethz.ch/handle/20.500.11850/286469 [Accessed 05 May 2019].

[2] IISD (2019). Tokenization of Infrastructure: A Blockchain-Based Solution to Financing Sustainable Infrastructure. International Institute for Sustainable Development, Winnipeg, Canada. Available at: https://www.iisd.org/sites/default/files/publications/tokenization-infrastructure-blockchain-solution.pdf [Accessed 05 May 2019].

[3] Verizon (2009). Data Breach Investigations Report. Verizon Business RISK Team, New York, NY. Available at: https://enterprise.verizon.com/resources/reports/2009_databreach_rp.pdf [Accessed 05 May 2019].

[4] DashLane (2018). Data Breaches 2018: The 20 Biggest Breaches of the Year. DashLane Inc., New York, NY. Available at: https://blog.dashlane.com/data-breaches-2018/ [Accessed 08 May 2019].

[5] First Data (2009). Data Encryption and Tokenization: An Innovative One-Two Punch to Increase Data Security and Reduce the Challenges of PCI DSS Compliance. First Data Corporation, Atlanta, GA. Available at: https://www.firstdata.com/downloads/thought-leadership/fd_encrypt_token_pci_whitepaper.pdf [Accessed 08 May 2019].

[6] Hargrave, J., Sahdev, N., and Feldmeier, O. (2018). How Value is Created in Tokenized Assets. SSRN Electronic Journal. DOI: 10.2139/ssrn.3146191.

[7] Deloitte (2018). The Tokenization of Assets is Disrupting the Financial Industry. Deloitte, London, UK. Available at: https://www2.deloitte.com/content/dam/Deloitte/lu/Documents/financial-services/lu-tokenization-of-assets-disrupting-financial-industry.pdf [Accessed 10 May 2019].

[8] Podolny, J.M., and Page, K.L. (1998). Network Forms of Organization. Annual Review of Sociology, 24: 57–76.

[9] Schollmeier, R. (2001). A Definition of Peer-to-Peer Networking for the Classification of Peer-to-Peer Architectures and Applications. Proceedings of the First International Conference on Peer-to-Peer Computing, 101–102. DOI: 10.1109/P2P.2001.990434.

[10] Morlino, L., and Quaranta, M. (2016). What is the Impact of the Economic Crisis on Democracy? Evidence from Europe. International Political Science Review, 37(5): 618–633.

[11] Hirsch, D.D. (2014). The Glass House Effect: Big Data, the New Oil, and the Power of Analogy. Maine Law Review, 66(2). Symposium: Who's Governing Privacy? Article 3.

[12] Nakamoto, S. (2008). Bitcoin: A Peer-to-Peer Electronic Cash System. Available at: https://bitcoin.org/bitcoin.pdf [Accessed 10 July 2019].
[13] Jentzsch, C. (2016), Decentralized Autonomous Organization to Automate Governance Final Draft—Under Review. Available at: https://download.slock.it/public/DAO/WhitePaper.pdf [Accessed 12 July 2019].
[14] Hacker, P. (2017). Corporate Governance for Complex Cryptocurrencies? A Framework for Stability and Decision Making in Blockchain-Based Organizations. Forthcoming in: Regulating Blockchain: Techno-Social and Legal Challenges. Oxford University Press, United Kingdom, 2019.
[15] Zuboff, S. (2019). The Age of Surveillance Capitalism: The Fight for a Human Future at the New Frontier of Power. Hachette Book Group, New York, NY.
[16] Brunton, F. (2019). Digital Cash: The Unknown History of the Anarchists, Utopians, and the Technologists Who Created Cryptocurrency. Princeton University Press, Princeton, NJ.
[17] Jos, P.H. (2006). Social Contract Theory: Implications for Professional Ethics. The American Review of Public Administration, 36(2): 139–155.
[18] Fukuyama, F. (2018). Contemporary Identity Politics and the Struggle for Recognition. Profile Books Ltd., London.
[19] Saran, S. (2018). A New Social Contract for the Digital Age. DOI: 10.13140/RG.2.2.15063.32169.
[20] Robbins, B.G. (2016). What is Trust? A Multidisciplinary Review, Critique, and Synthesis. Sociology Compass, 10(10): 972–986. DOI: 10.1111/soc4.12391.
[21] The Great Hack (2019). Available at: https://www.netflix.com/Title/80117542 [Accessed 26 July 2019].
[22] Isaak, J., and Hanna, M.J. (2018). User Data Privacy: Meta, Cambridge Analytica, and the Privacy Protection. Computer, 51(8): 56–59.
[23] Okenyi, P.O., and Owens, T.J. (2007). On the Anatomy of Human Hacking. Information Systems Security, 16: 302–314.
[24] Hobbes, T. (2010). Leviathan: Or The Matter, Forme, and Power of a Common-Wealth Ecclesiasticall and Civill. Cambridge University Press, Cambridge.
[25] Tyler, T. (2006). Why People Obey the Law. Princeton University Press, Princeton.
[26] IETF (2014). On Consensus and Humming in the IETF. Available at: https://tools.ietf.org/html/rfc7282 [Accessed 17 July 2019].

[27] Liu, L., and Xiong, L. (2004). Peer Trust: Supporting Reputation-Based Trust for Peer-to-Peer Electronic Communities. IEEE Transactions on Knowledge and Data Engineering, 16.

[28] Werbach, K. (2018). The Blockchain and New Architecture of Trust. The MIT Press, Cambridge, MA.

[29] Adam, I., and Fazekas, M. (2018). Are Emerging Technologies Helping Win the Fight against Corruption in Developing Countries. Available at: http://www.govtransparency.eu/wp-content/uploads/2019/02/ICT-corruption-24Feb19_FINAL.pdf [Accessed 16 July 2019].

[30] Data Protection (2018). Rules for the Protection of Personal Data Inside and Outside the EU. Available at: https://ec.europa.eu/info/law/law-topic/data-protection_en [Accessed 19 July 2019].

[31] Blockchain-Consortium (2018). How Does the EU's GDPR Apply to Hashed Data on the Blockchain? Available at: https://legalconsortium.org/uncategorized/how-does-the-eus-gdpr-view-hashed-data-on-the-blockchain/ [Accessed 20 July 2019].

[32] Qedit (2019). A Privacy Layer for Your Blockchain Network. Available at: https://qed-it.com/ [Accessed 20 July 2019].

[33] ZKProof (2019). Zero Knowledge Proof Standardization. Available at: https://zkproof.org/ [Accessed 20 July 2019].

[34] Shrobe, H.E., Shrier, D.L., and Pentland, A. (2017). New Solutions for Cybersecurity. Cambridge, MA: The MIT Press.

[35] Solid (2019). How Solid Realizes the Web as It Was Originally Envisioned, in Layman's Terms. Available at: https://solid.inrupt.com/how-it-works [Accessed 20 July 2019].

[36] EPFL (2019). DEDIS. Available at: https://dedis.epfl.ch/ [Accessed 25 July 2019].

[37] Elixxir (2019). Introduction. Available at: https://elixxir.io/introduction [Accessed 25 July 2019].

[38] Fenwick, M., and Vermerlen, E.P.M. (2018). Technology and Corporate Governance: Blockchain, Crypto, and Artificial Intelligence. Lex Research Topics in Corporate Law & Economics Working Paper, No. 2018-7.

[39] Zagar, T.M. (2018). A New Chapter for ICONOMI: Transformation of Corporate Governance and Issuance of Equity Tokens. Available at: https://medium.com/iconominet/a-new-chapter-for-iconomi-transformation-of-corporate-governance-and-issuance-of-equity-tokens-dc603df2272b [Accessed 25 July 2019].

[40] Blemus, S., and Guegan, D. (2019). Initial Crypto-Asset Offerings (ICOs), Tokenization, and Corporate Governance. Available at: https://ssrn.com/abstract=3350771 [Accessed 26 July 2019].

[41] Libra (2019a). White Paper. Available at: https://libra.org/en-US/white-paper/ [Accessed 20 July 2019].
[42] Libra (2019b). Section 05 The Libra Association. White Paper. Available at: https://libra.org/en-US/white-paper/#the-libra-currency-and-reserve [Accessed 20 July 2019].
[43] Evans, B.J. (2011). Much Ado About Data Ownership. Harvard Journal of Law & Technology, 25(1): 69–130.
[44] Zyskind, G., Nathan, O., and Pentland, A. (2015). Decentralizing Privacy: Using Blockchain to Protect Personal Data. MIT Media Lab. Available at: https://web.media.mit.edu/~guyzys/data/ZNP15.pdf [Accessed 28 May 2019].
[45] Ellen-MacArthur Foundation (2013). Towards the Circular Economy: Economic and Business Rationale for an Accelerated Transition. Available at: https://www.ellenmacarthurfoundation.org/assets/downloads/publications/Ellen-MacArthur-Foundation-Towards-the-Circular-Economy-vol.1.pdf [Accessed 28 May 2019].
[46] Centre of Expertise on Resources (2019). The Circular Economy and Developing Countries: A Data Analysis of the Impact of a Circular Economy on Resource-Dependent Developing Nations. COE-Resources Issue Brief 3.
[47] RecycleToCoin (2019). Introduction. Available at: https://www.recycletocoin.com/ [Accessed 30 May 2019].
[48] GiftPay (2019). Solutions. Available at: https://www.giftpay.co.uk/business/solutions.aspx [Accessed 30 May 2019].
[49] PlasticBank (2019). What We Do. Available at: https://www.plasticbank.com/what-we-do/ [Accessed 30 May 2019].
[50] IBM LinuxONE (2019). IBM LinuxOne. Available at: https://www.ibm.com/it-infrastructure/linuxone [Accessed 30 May 2019].
[51] United Kingdom Parliament (2016). Food Waste: Key Facts, Policy, and Trends in the UK. Available at: https://researchbriefings.parliament.uk/ResearchBriefing/Summary/CBP-7552 [Accessed 14 July 2019].
[52] Goodr (2019). Feed More: Waste Less. Available at: https://goodr.co/about-us/ [Accessed 14 July 2019].
[53] Goldstein Research (2019). Global Tire Recycling Market Analysis: Opportunity, Demand, Growth, and Forecast 2016–2024. Available at: https://www.goldsteinresearch.com/report/global-tire-recycling-industry-market-trends-analysis [Accessed 14 July 2019].
[54] EV Volumes (2019). Global EV Sales for 2018—Final Results. Available at: http://www.ev-volumes.com/country/total-world-plug-in-vehicle-volumes/ [Accessed 23 July 2019].

[55] PowerLedger (2019). Power Ledger and Silicon Valley Power Trial to Turn Electric Vehicles into Mobile ATMs. Available at: https://www.powerledger.io/article/power-ledger-and-silicon-valley-power-trial-to-turn-electric-vehicles-into-mobile-atms/ [Accessed 17 July 2019].

[56] Snell, R. (2019). The Rise of Autonomous Vehicles and Why Ethics Matter. Available at: https://www.digitalistmag.com/improving-lives/2019/04/02/rise-of-autonomous-vehicles-why-ethics-matter-06197534 [Accessed 18 July 2019].

[57] Cheney-Lippold, J. (2017). We Are Data: Algorithms and the Making of Our Digital Selves. New York University Press, New York, NY.

PART III

The Humanized Internet Enablers

DOI: 10.1201/9781003339724-8

5

Ethics, Governance, and Regulatory Challenges

5.1 Introduction

Can we imagine the abuses of facial recognition and any biometric use of data for our identity? There is not a lack of these abuses and certainly what may be interpreted as an abuse can be defined as a state surveillance or perhaps a new definition of societal harmonization. Within the context of societal dilemmas, how we define and agree to biometric use of data is daunting and can quickly lead to a very dystopian narrative [1].

Biometric-based Digital Identity Initiatives have been emerging in Africa and India, ID4Africa and the Aadhaar project for the past several years. These national initiatives fall under the UN Sustainable Goal, 16.9, for legal identity for all by 2030. These projects have not been without challenges; however, we do note that national identity initiatives are commencing with biometric mechanisms. What we need to postulate is how far can a national biometric database go e.g. your DNA and understand the potential for abuse [2].

What about behavioral biometrics for eye tracking, keystroke pattern development and so on to develop a profile about you or me?

Your political preferences and sexual identity can be ascertained with you knowing it [3].

What we observe is that privacy is no longer about data compliance but rather a human right. When that right is egregiously breached, what course of redress do you or I have?

Clicking that pervasive "Like" button only exacerbates the situation further in terms of ascertaining the YOU, YOUR predilections and monetizing your online persona further.

Can we assert that there is a weaponization of our internet usage where the perpetrators could be unknown by the victims and, the boundaries are borderless?

DOI: 10.1201/9781003339724-9

What do we mean by ethics and governance in this brave new world where technology is weaponized to either control a citizen and/or hurt an individual?

The key narrative of this chapter is to explore foundationally why a framework of ethics, governance and regulation is critical to our society.

We require a multi-stakeholder approach to contextually define ethics, governance, and regulation.

In the end, how do we define "responsible use of technology" such that all multi-stakeholders are able to embrace the definition and moreover act on it?

Further, we need a new approach to redress digital harms to either an individual and/or to an organization.

In Chapter 5, we explore what has gone awry with current systems prompting an overview that describes the Human Enablers. Proposing a "Digital New Deal" in defining a framework for Ethics, Governance and Regulation is an opportunity here.

5.2 The Current Systems Are No Longer Tenable

The current systems as we know there are no longer tenable. Why? There are many questions we can pose that cross a wide spectrum for our societal dilemma under the responsible use of technology.

What are the systemic issues that are driving issues of mis/disinformation? What are the root causes of those issues and the symptoms of them? How does this ultimately extend to an industry-wide issue? A society-level issue? A humanity-level issue? Who is winning and who is losing?

What we mean by "system" in this discussion is highly contextual in terms of culture and jurisdiction to name a few examples. When thinking of the internet, a system evokes interconnections as in neural networks. We use the internet and related applications to communicate. The internet simply works, or does it? The Internet Engineering Task Force or IETF mantra of rough consensus and working code has required clarification that is captured in RFC 7282 [4]. A system can also refer to a government entity and herein lies the conundrum. What is a lie can be truth and the development of social media platforms amplifying "alternative truths" whether disinformation or misinformation presents us with the narrative that systems as we know them are no longer tenable.

The Digital Pandora's box is now open and has been over these past several years. When Ursula von der Leyen the President of the European Commission asserts that Europe wants 'algorithms that protect people" or "human center digital transformation that protects identity and privacy," what does she mean? [5]

Or when an EU plane is hijacked because a blogger who is determined to be a terrorist by the hijacking entity occurs, where does one go for jurisdictional correction? [6]

Weaponization of social media and the damage it causes in reputation, or far worse, resulting in suicide only amplifies why systems as we know them no longer suffice.

"Live by the digit, die by the digit" in the form of cyberbullying is our reality.

Determining "responsible use" in partnership with governments and private industry is a step forward, e.g., a digital Hippocratic oath. However, when the red line has been crossed and, harm has ensued, who does one call today? Humans create these technologies and we really need to define the intentional use, especially as the teams that develop the tech may not reflect who the beneficiaries of these solutions are. The implication is a proliferation of unwanted bias.

This observation begs the question as to how this chasm may be resolved, and by which entities? What kind of social contract is required for the digital world we share?

The 21st century, fostered by the use of social media and the internet affirms that social contracts have been evolving to identity politics and the notion of multiple identities especially when referring to the "Internet of Everything and Everyone." Even worse is the trajectory toward dystopian control.

In the provocative book, *Digital Transformation and Ethics* by Peter G. Kirchschlerger [7] describes the sense of urgency to respond to these scenarios.

"There is an urgency to act. 'Surveillance capitalism' is not the only possible way; there are alternatives. The current Chinese government gives humanity an outlook on what the entire world could look like if we don't act immediately... The current Chinese government is introducing – with support of multinational technology corporations – a social points system throughout its entire territory consisting in the evaluation of humans according to their social credit...This total surveillance by a dictatorship is combined with the biotechnological and generic breaking of taboos."

In their book, *An Ugly Truth, Inside Meta's Battle for Domination*, [8] authors Sheera Frenkel and Cecilia Kang describe the history of Meta and its founder Mark Zuckerberg. The pivots to protect the business at all cost are chronicled in this book. Early on while at Harvard, the question of privacy had surfaced.

"Zuckerberg's site assured students that because the network was limited to Harvard, it was private by design. But Meta's earliest terms of service made no mention of how users' personal details (which they hadn't come to think of as their individual data) could be used… In one online chat, he made clear just how much access he had to the data he had accumulated. Zuckerberg began the conversation with a boast, telling one friend that if he ever needed information on anyone at Harvard, he should just say the word:

> ZUCK: I have over 4000 emails, pictures, addresses, sms
> FRIEND: what!? How'd you manage that one
> ZUCK: people just submitted it
> ZUCK: I don't know why
> ZUCK: they "trust me"
> ZUCK: dumb fucks"

In hindsight, we know that social media has been used by foreign governments to influence election outcomes. The pattern of "monetized algorithmic truth decay" has been evolving over sometime now. If we live in filter bubbles that amplify truths that are contextual, are we safe?

Think for example spread of horse deworming drug, Ivermectin as treatment for COVID-19 amplified over the TikTok platform [9].

"I think it's the worst (platform) by far, actually, at curbing misinformation," Dr. Mrozinski says of TikTok. "Because of the way the algorithm works, and the For You page, which is primarily what people flip through as they go. And it's short-form, so the videos that do quite well are 10 to 15 seconds and have a really punchy hook and get to the point quite quickly. And it's easily shared as well, and gets pushed out to more and more people." In the United States, the use of Ivermectin and the right of citizens to select the drug as an alternative treatment for Covid-19 has been further politicized [10].

> "Republicans have apparently taken it upon themselves to ensure that Americans continue to receive terrible medical advice now that **Donald Trump** is no longer offering it up from the White House podium. In the spirit of the former president—whose own penchant for unproven and often dangerous COVID-19 treatments has included suggesting Americans inject bleach or expose themselves to "very powerful light"—Senator **Ron Johnson** has for months supplemented his vaccine conspiracy theories with enthusiasm for unsound COVID-19 treatments. Most recently, he's latched onto ivermectin, a deworming drug intended for horses and other livestock, and with the assistance

of conservative media, helped prompt a warning from the FDA after people were hospitalized for taking it. The anti-parasitic is approved for certain uses in humans, such as head lice, but it "has repeatedly failed in clinical trials to help people infected with the coronavirus," according to the *New York Times*."

These examples (and many more) only amplify the fact that the systems as we know them are broken and require a new approach perhaps in the form of a framework for Ethics. Governance and Regulation which we propose in the next section.

5.3 Defining a New Framework for Ethics, Governance, and Regulation

Author Susan Liautaud, describes in her book, *The Power of Ethics* [11] that "compromised truth degrades both memory and history." Susan Liautaud asserts further that "compromised truth is binary – an unacceptable humanity – defining and humanity destroying condition....Compromised truth destroys our identity. The endgame, then, is this: In a post-truth world, we cannot see what we are seeing anymore... We lose all reference points – times, places, people, decisions. Compromised truth is like ethical quicksand that swallows us up completely."

It is no surprise that there is a struggle in describing what we mean by ethics in technology. Whose ethics? How does one avoid cognitive bias? What does the composition of the developer team look like? Therefore, embedding ethics in technology could manifest itself in responding to questions like identifying the intentional use of the technology itself. Outlining potential for abuse is an opportunity here for technologists. The sum of these actions may translate to overall safety for consumers and serve as guidance for policymakers. Make no mistake this is not an easy task but having platforms regulate themselves is no longer viable. Education and training will be foundational to this process whether in universities or in startup communities themselves. At a corporate level, what will be essential is to understand the red lines that impact the organization's brand for example u articulating what is meant by corporate accountability and governance. Make no mistake, corporations are beholden to their shareholders first and foremost. Perhaps, having a corporate narrative in this area of technical ethics and governance that extends beyond the typical legal ethics document one often sees is an opportunity for corporations. Finally, it is incumbent upon everyone: consumers, corporations, and technology developers to pose questions as to potential misuse of technology and course of action to mitigate abuse. The debate continues!

5.4 Summary

In Chapter 5, double-clicking on ethics, governance and regulatory challenges produce more questions than answers. These themes are contextual. However, one common thread is the possibility of abuse. Have we become numb to what could be misuse of these technologies? Let's face it, when "bad stuff" happens to you the narrative is different. The point is that consumers specifically must be the center of this discussion and to a certain extent be in "control."

In Chapter 6, we will explore standardization and in Chapter 7 the notion of ethical machines.

References

[1] https://www.npr.org/2021/01/05/953515627/facial-recognition-and-beyond-journalist-ventures-inside-chinas-surveillance-sta

[2] https://www.biometricupdate.com/id-for-all

[3] https://www.biometricupdate.com/biometric-news/behavioral-biometric-news-articles

[4] https://datatracker.ietf.org/doc/html/rfc7282

[5] https://leading-the-digital-decade.eu/

[6] https://www.theatlantic.com/ideas/archive/2021/05/belarus-ryanair-hijacking/619028/

[7] Digital Transformation and Ethics, Ethical Considerations on the Robotization and Automation of Society and the Economy and the Use of Artificial Intelligence, Peter G. Kirchschlaeger, Nomos 2021.

[8] An Ugly Truth, Sheera Frenkel and Cecilia Kang, The Bridge Street Press, 2021.

[9] https://www.rollingstone.com/culture/culture-news/tiktok-covid-cures-ivermectin-horse-deworming-medication-1216268/

[10] https://www.vanityfair.com/news/2021/08/republicans-bravely-defend-americans-rights-to-treat-covid-with-horse-dewormer

[11] The Power of Ethics, How To Make Good Choices in a Complicated World, Susa Liautaud, Simon&Schuster, 2021.

6

Standardization

6.1 Introduction

In this chapter, we explore the Standardization landscape that intersects with identity and technology ethics. The question we must all pose is what is meant by the development of standards within these two areas given the fact that these topics are contextual in nature?

Identity is multimodal includes culture. Decentralization of identity and the related standards activities are the purview of the Decentralized Identity Foundation. [1] The Sovrin Organization has had to re-define itself specific to accountability and governance. [2] The World Economic Forum (WEF) has recently created a Digital Identity Impact Working Group under the umbrella of its Crypto Impact and Accelerator Program or CISA. One of the authors of this book is active in CISA and in this working group. [3] WEF's primary narrative for digital identity is that we have been witnessing an evolution these past several years. self-sovereign identity (SSI) is a subset of this discussion core to the ownership of personal data. "Things" possess identities (Internet of Things for example) like wearables; sensors; smart grid; smart cars, smart cities and so on. [4] Internet of Things or IoT identities possess their own set of challenges mainly in the cybersecurity space. [5] With the development of the metaverse or metaverses, standardization is also a key topic within the context of privacy.

> "If in the metaverse users are the custodians of their own data and digital identity, what, if any, rules and regulations will apply to them? How do they maintain and protect their data? What recourse would they have if some or all of it, even their entire digital identity itself is stolen, modified, or cloned?"

DOI: 10.1201/9781003339724-10

Alex Pentland of MIT, in the December 2022 Communications of the ACM viewpoint article entitled, “Building a New Economy: Data, AI and Web3, How Distributed Technologies Could Return More Control to Users” asserts that “there is an urgent need to develop standards that guarantee a Web3 economy that remains truly distributed and provides global interoperability along with adequate protection for individuals and communities.” [6]

The Institute of Electrical and Electronic Engineers or IEEE [7] is active across the spectrum of the discussion regarding ethics overall. One of the authors of this book is the chair of the IEEE Global Ethics in Extending Reality Working Group. [8] She is also working across IEEE with colleagues to further develop an ethical component for metaverse standards development. [9]

Let us explore a little further the standards landscape for identity and technology ethics in the next sections of this chapter.

6.2 Decentralized Identity Foundation (DIF)

The Decentralized Foundation or DIF describes itself as an “engineering driven” organization and includes working groups such as Identifiers and Discovery; Decentralized Identifiers or DIDs. [10] The World Wide Web Consortium (W3C) announced on July 19, 2022, that DIDs v.1.0 is a W3C recommendation. [11] The development of technical specifications for decentralized identity has been the focus of DIF. The premise for DIF is to develop an “identity layer” for the internet. There are Special Interest Groups or SIGs like “Travel and Hospitality” [12] or “Finance and Banking.” Specific to Europe is the creation of the European Digital Identity Framework [13] and the EU Digital Wallet Consortium [14] for the EU population of circa 450 million citizens.

A word about self-sovereign identity (SSI) is best captured in a September 26, 2022, Forbes article entitled, “Self-Sovereign Identity: Taking Control Over Your Digital Identity,” written by Victor Shilo that advocates that the individual must be in control of what information to share and when.

> “The central tenet of SSI – with its 10 guiding principles that Christopher Allen defined in 2016 – is that everyone has a *moral* right to be in control of their identity. Each of us decides whether to share private information with any entity for authentication as well as *what* to share and *when* to share it. No silos of personal data are created. In other words,

> both attributes and credentials, which can be self-signed or signed by third parties, must be controlled by users who present their data to service providers (SPs) whenever they desire. [15]

With SSI, a secure peer-to-peer channel is established between the ID issuer, the ID owner and the ID verifier. The ID owner chooses which attributes of their identity they want to "show" to the ID verifier."

There are two key components of SSO: 1. Verifiable Credentials (VCs) and 2. Decentralized Identifiers (DIDs). In the same article, the author illustrates the following concrete example:

> "*Imagine the government wants to issue you a digital version of your passport alongside your physical copy. At the municipal service desk, you are asked to scan a QR-code. Here, the DIDs are exchanged, creating a secure connection. Through this secure connection, the clerk now issues you your digital passport in the form of a Verifiable Credential. You accept, storing it in your (digital) wallet.*"

The corpus of work that DIF has progressed coupled with relevant use cases continue to evolve. Standardization therefore translates to industry adoption. DIF is an example of balancing between research and pragmatic approach to generating use cases that may delight enterprises and consumers.

6.3 Institute of Electrical and Electronic Engineers (IEEE)

IEEE is quite active in undertaking "ethics" the development of standards under the Society for Social Implications of Technology Standards Committee (SSIT). [16] The remit of IEEE SSIT is as follows:

SSIT is developing standards to help guide stakeholders through the complex interactions of technology development and deployment with society, in order to ensure that we all benefit from technology.

SSIT focuses on five major areas:

- Sustainable Development & Humanitarian Technology
- Ethics, Human Values and Technology
- Technology Benefits for All
- Future Societal Impact of Technology Advances
- Protecting the Planet & Sustainable Technology" [17]

One of this book's authors is an IEEE Senior Member and is the Chair for Ethics in Extended Reality Standards Group (IEEE P7030). As captured in the IEEE P7030 Working Group Scope:

> "This standard establishes a uniform set of definitions, and a methodology to assess the socio-technical considerations and practices regarding ("XR") Extended Reality (Augmented Reality, Virtual Reality, Immersive Web and Spatial Web technologies) where this methodology shapes the positive design of XR systems"

The participants come from eclectic backgrounds and are invited to contribute their knowledge into this evolving standard. [18]

SSIT Standards chair, Ruth Lewis, invited a group of her group's chairs including one of the authors of this book to a November 10, 2022, Standards Track under the IEEE International Symposium of Technology and Society or ISTAS Conference to amplify their work and status. [19] IEEE ISTAS 2022 included an Ethics Day session. [20]

The IEEE standards development under "Ethics and SSIT" is just the tip of the iceberg and an example of how the creation of standards that is inclusive.

6.4 Summary

In this chapter, we have taken a snapshot of the Standards Landscape for Identity and Ethics in Technology. One common theme is the parallel creation of both standards and industry use cases designed to preempt fragmentation and harmonize standards frameworks that may be interoperable. This observation is not easy as the modus operandi is often "let the market decide." At some point, and perhaps an area for further research is to assess the cost of standards development, and implied market fragmentation to the actual benefits. There is a polarity between waiting for standards and stimulating the market (the latter often taken by private enterprises).

The recommendation is to be more pragmatic.

In Chapter 7 we will explore "Ethical Machines "and to discuss the embodiment of what ethics can look like. As previously, articulated, the potential for harm exists. The question is whether we are able to invent ethical machines and again whose ethics? Does the human remain in the loop or removed?

References

[1] https://identity.foundation/governance/about
[2] https://sovrin.org/announcement-enforcement-of-accountability-requirements-on-the-sovrin-networks-by-1-february-2023/
[3] https://initiatives.weforum.org/crypto-impact-and-sustainability-accelerator/home
[4] https://www.weforum.org/agenda/digital-identity/
[5] https://www.weforum.org/agenda/2021/03/ai-is-fusing-with-the-internet-of-things-to-create-new-technology-innovations
[6] https://cacm.acm.org/magazines/2022/12/266920-building-a-new-economy/abstract
[7] https://www.ieee.org/
[8] https://sagroups.ieee.org/7030/
[9] https://standards.ieee.org/beyond-standards/industry/technology-industry/why-are-standards-important-for-the-metaverse/
[10] https://identity.foundation/
[11] https://www.w3.org/2022/07/pressrelease-did-rec.html.en
[12] https://www.phocuswire.com/phocuswright-europe-ssi-ddid-resources
[13] https://ec.europa.eu/commission/presscorner/detail/en/ip_21_2663
[14] https://eudiwalletconsortium.org/
[15] https://www.forbes.com/sites/forbestechcouncil/2022/09/26/self-sovereign-identity-taking-control-over-your-digital-identity/?sh=6c33549664e0
[16] https://sagroups.ieee.org/ssit/
[17] Ibid.
[18] https://sagroups.ieee.org/7030/
[19] https://www.istas22.org/program/track-10/#s1
[20] https://www.istas22.org/program/ethics-day/

7

Ethical Machines

7.1 Introduction

As our world transforms digitally accelerated by technologies like AI, blockchain, quantum computing, and so on, one has to ponder the results of these technologies in terms of machines themselves. We have established that there is a polarity between creating services that will be both beneficial and transparent for all, to creating services that have the potential for consumer abuse examples like “intelligent” finance; “intelligent” healthcare; “intelligent” travel and hospitality all in the form of “benevolent” profiling. Profiling occurs predictively given past behavior as inputs and provided to companies that want to monetize these behaviors. The new AI-generated BOT, ChatGPT [1] may have its challenges but the Pandora’s box has been opened. [2] In recent years, the rapid advancements in artificial intelligence have revolutionized industries, empowered innovation, and raised profound ethical questions about the role of machines in our society. As AI systems continue to evolve and exhibit levels of autonomy not seen before, the importance of instilling ethical principles into these machines has gained significant traction. The concept of ethical machines, or AI systems programmed to operate in alignment with ethical norms and values, has emerged as a crucial area of focus in the field of AI ethics. There is even movement to “pause” the use of ChatGPT with Italy temporarily banning its use due to privacy concerns. [3] Further, the “Godfather of AI, Dr. Geoffrey Hinton, has warned of the “growing dangers from developments in the field,” asserting that AI chatbots were “quite scary.” [4] Open AI CEO (ChatGPT) Sam Altman, has stated that that “artificial intelligence should be regulated by a US of Global Agency,” that the concern about ChatGPT’s use “has expanded to broader concerns about the ability if the latest ‘generative AI’ tools to mislead people, spread falsehood, violate copyright protections, and upend some jobs.” [5]

DOI: 10.1201/9781003339724-11

On June 14, 2023, the "European Parliament approved "a regulation of the European Parliament and of the Council on laying down harmonized rules on artificial intelligence (Artificial Intelligence Act) and amending certain Union legislative acts (COM (2021)0206 – C9-0146/2021 – 2021/0106(COD))"

The text of the approved regulation captures the spirit of the regulation itself as follows:

> "The purpose of this Regulation is to promote the uptake of human centric and trustworthy artificial intelligence and to ensure a high level of protection of health, safety, fundamental rights, democracy, and rule of law and the environment from harmful effects of artificial intelligence systems in the Union while supporting innovation and improving the functioning of the internal market." [6]

While you are sleeping these interactions are happening and will have impact to our society.

The question is whether you are aware. Do you care?

The notion of "ethical machines" provides us with an opportunity to define the "ethical machine" moreover to develop an "*Ethical Machine Manifesto*" for the industry.

In this chapter, we delve into the intersection of ethical machines and the concept of the Humanized Internet. We explore the significance of AI systems that align with ethical principles and values and how they could contribute to creating a more inclusive, respectful, and digital landscape. By considering the ethical implications of technology and promoting human-centric approaches, we can shape an internet that upholds human dignity, digital identity, and democracy. The present and the future do not have to be dystopian if only we are aware of technology consequences that extend beyond the commercialization of technology, but rather understand the potential for misuse and the benefits.

Quo Vadis the "ethical machine?"

7.2 The Ethical Machine

In the 2017 book entitled, "*What To Do When Machines Do Everything, How to Get Ahead in a World of AI, Algorithms, BOTs, and Big Data*," the authors Malcolm Frank, Paul Roehrig and Ben Pring, define "the New Machine" as "a system of intelligence combines *software* (algorithms, business rules, machine-learning code, predictive analytics), *hardware* (servers, sensors, mobile devices, connectivity), *data* (contextualized and real time), and *human input* (often judgment or questions)." [7] The essential thesis is not to worry about ethics but rather to focus on the value-add proposition for the industry specifically for companies. Under "Data is the New Oil" moniker, with all the data that we pulsate daily, this observation only confirms the thesis. However, rather than condemning such viewpoints, it is essential to understand the evolution that has been transpiring and certainly the implications to our society.

In the book, "*Algorithms to Live By, The Computer Science of Human Decisions*," published in 2016, the authors Brian Christian and Tom Griffiths, under "Idolatry of Data" state that "If we had copious data, drawn from a perfectly representative sample, completely mistake-free, and representing exactly what we are trying to evaluate, then using the most complex model available would be the best approach. But if we try to perfectly fit our model to the data when any of these factors fails to hold, we risk overfitting. In other works, overfitting poses a danger over time when we are dealing with noise or mismeasurement." [8] The context is how we can implement algorithmic thinking to drive our own daily routines. The implication is that we may very well feed inputs into a "machine" as the human in the loop. What "machine" and by whom? The "what" and "who" are rather nebulous. After all, we are humans and "human-centric design" is core to this narrative toward the development of ethical machines. [9]

By its very name, "ethical machine," one could assume that there is a notion that the ethical machine or perhaps, more accurately, the "ethical AI system" is designed to operate in accordance with ethical principles and guidelines. The "ethical AI system" is developed with a focus on fairness, transparency, accountability, and the well-being of individuals and society. The author of this book uses the terms, "ethical machine" and "ethical AI System" interchangeably throughout this chapter as to indicate the importance of embedding ethics in the design process for algorithms. There is a corpus of publications written about ethical machine constructs ranging from the book, *Machine Ethics*, which is a compendium of essay that focuses on ethical research with Michael Anderson and Susan Leigh Anderson as editors or *Ethical Machines*, [11] written by Reid Blackman which is a practical guide in implementing "unbiased, transparent and respectful AI," with business organizations as the key target audience. MIT has implemented the moral machine platform for "gathering a human perspective on moral decisions made by machine intelligence. [12] These publications should be reviewed, and the research followed for a foundational overview in this space. The development of ethical machines involves integrating ethical frameworks and principles into the design and programming process. Examples of principles include human rights, non-discrimination, consent, and the avoidance of harm (recall the Hippocratic oath, "Do No Harm"). The ethical machine must be adaptable and responsive to societal permutations and cultural contexts, as ethical standards can vary across different communities and time periods.

In recent years, the rapid advancement of artificial intelligence (AI) has brought forth numerous ethical dilemmas and challenges, As machines become increasingly intelligent and autonomous, it is imperative to consider the ethical implications of their actions.

The goal of ethical machines is to bridge the gap between technological advancements and ethical considerations between technological advancements and ethical considerations, ensuring that AI systems and robots contribute positively to society without compromising human values or causing harm. (Note once again inference to "Do no harm"). By promoting responsible and ethically aligned AI, ethical machines aim to address concerns related to the impact of automation and intelligent systems on various aspects of human like employment, privacy, and social interactions.

Who creates ethical machines?

Ethical machines are typically created through a collaborative effort involving various stakeholders, including researchers, engineers, ethicists, policy makers and industry professionals. Let's explore some key players involved in the creation of ethical machines:

1. **Researchers and engineers:** Experts in fields such as artificial intelligence (AI), machine learning, robotics, and computer science play a vital role in developing the technical aspects of ethical machines. They design algorithms, develop models, and build systems that adhere to ethical principles.

2. **Ethicists and philosophers:** Ethicists and philosophers provide ethical guidance and frameworks for the development of ethical machines. They analyze the ethical implications of AI technologies, contribute to the design of ethical decision-making processes, and ensure that ethical principles and values are integrated into the development process.

3. **Policymakers and regulatory bodies:** Governments and regulatory bodies play a crucial role in shaping the development and deployment of ethical machines. They establish legal and ethical frameworks, guidelines, and standards to ensure that AI technologies adhere to ethical considerations and societal values. They may establish regulations for data protection, privacy and transparency.

   Though the role that policymakers and regulatory bodies assumes is a multi-stakeholder approach, perhaps one could envisage a policymaker or regulator who is technology savvy, and may have a background in computer science, law, and ethics. This background and a priori experience should be balanced with multistakeholder consultations.

Ethical machines represent a pivotal frontier in AI development, requiring comprehensive research efforts to ensure the responsible and ethical deployment of AI systems. By addressing key research needs in transparency, bias mitigation, accountability, and human-AI collaboration, we can navigate the complex landscape of AI ethics and security to build a more inclusive and trustworthy AI-powered future. At its core, the concept of ethical machines revolves around imbuing artificial intelligence systems with the capacity to make decisions and take actions that align with ethical principles and societal values.

While machines lack human-like consciousness, they can be designed to adhere to predefined ethical guidelines and norms, thus enabling them to operate in a manner that upholds ethical standards. Ethical machines refer to AI systems that exhibit ethical behavior, considering social norms, moral values, and legal requirements in their decision-making processes. The importance of ethical machines lies in their potential to mitigate harm, promote fairness, and enhance trust in AI technologies. By incorporating ethical considerations into AI design and development, we can navigate complex ethical dilemmas and ensure that AI systems are aligned with human values.

The notion of ethical machines traces back to the early discussions on AI ethics and the need to imbue AI systems with ethical reasoning capabilities. Over the years, researchers, ethicists, and policymakers have grappled with the challenges of developing AI systems that not only perform tasks efficiently but also ethically. From Asimov's Three Laws of Robotics to contemporary ethical frameworks such as the IEEE Global Initiative for Ethical Considerations in AI and Autonomous Systems, ethical machines have been a central theme in shaping the ethical landscape of AI.

In the current landscape, ethical AI is a burgeoning field that encompasses diverse perspectives on how AI systems should be designed, deployed, and governed to uphold ethical standards. Initiatives focusing on transparency, fairness, accountability, and explainability have gained traction, aiming to address ethical challenges in AI development. However, as AI technologies advance, the need for robust ethical frameworks and proactive measures to ensure ethical machine behavior becomes increasingly pressing.

4. **Industry and tech companies:** Technology companies and industry players are responsible for the creation and implementation of AI systems, including ethical machines. They invest in research and development, collaborate with experts, and incorporate ethical considerations into the design and deployment of their technologies. Companies also have a responsibility to address ethical challenges and ensure that their AI products and services meet ethical standards**.** Trust is foundational here. The polarity between revenue-generating opportunities and ethics is a factor. However, the measure of how important ethics in AI may be seen by where the topic is within a company e.g. the establishment of AI ethics officer that reports directly to the CEO. A company like eqtylab.io has developed tools to foster the role of ethics and trust in AI see: https:// www.eqtylab.io

   More the development of trust tools for AI a core tenet to the company.

5. **Policy makers:** In fact, companies have struggled in this space calling out more for regulation or define the guardrails clearly. We are commencing to see examples of government regulation and guidelines. Governments play a crucial role in legislating ethical practices in AI for example. Establishing frameworks for accountability, transparency and fairness is essential. Regulatory bodies should monitor AI deployment and impose guidelines that ensure ethical considerations are integrated into technological development. The European Union (EU) has initiated the AI Act, "The AI Act is the first-ever legal framework on AI, which addresses the risks of AI and positions Europe to play a leading role globally." (https://digital-strategy.ec.europa.eu/en/policies/regulatory-framework-ai)

The US National Institute of Standards and Technology (NIST) is pursuing signed agreements for AI Safety research. https://www.nist.gov/news-events/news/2024/08/us-ai-safety-institute-signs-agreements-regarding-ai-safety-research

As technological advancements proliferate, particularly in AI and machine learning (ML), the introduction of machines capable of ethical decision-making has become paramount. The interplay between ethics and machine functionality poses a complex dilemma that necessitates clarity on where moral responsibilities lie. Ethical machines fundamentally are systems designed to behave in ways that align with human morals and societal ethical standards (contextual). These machines utilize algorithms that integrate ethical frameworks, assisting them to make decisions that are in the best interest of individuals and society. An example that is commonly cited are autonomous vehicles that can analyze potential accident scenarios and consequently must determine the least harmful outcome for all involved. One of the paramount reasons for developing ethical machines is to ensure the safety of their users ("Do no harm").

As previously stated, one of the paramount reasons for developing ethical machines is to ensure the safety of their users and the broader community. Unregulated AI systems can pose risks that may lead to catastrophic consequences If they make harmful decisions. For instance, AI used in healthcare must weigh the benefits and risks of treatment options, ideally prioritizing patient safety and well-being. Building trust is so foundational to this narrative. Ethical machines must foster public trust in technology. As machines increasingly influence everyday life, from healthcare diagnostics to law enforcement, society must trust that these systems will make unbiased, fair and just decisions. Trust in turn promotes the adoption of AI solutions. Ethical machines provide transparency and accountability reassuring users about the integrity of algorithms used in decision-making processes.

In reducing bias and ensuring fairness, we must note that human biases often seep into AI through data, resulting in unjust outcomes based on race, gender and socioeconomic status. Ethical machines that actively seek to mitigate bias are critical in creating a more equitable society. Reinforcing fairness mechanisms in AI systems ensure that marginalized communities receive equitable treatment that promotes social justice through technological advancements. The allocation of responsibilities like responding to the question of who is accountable for the actions of ethical machines is complex and multi-faceted. The responsibilities are shared amongst various stakeholders ranging from the company CEO; developers; users, and governments.

Developers and technologists, the context here is the designers of AI algorithms. They must prioritize ethical considerations during the development process and employ inclusive practices to mitigate bias. Continuous education on ethics in AI is essential to emphasize the significance of human-centered design principles.

Users or consumers of AI must engage with these systems responsibly and critically. They should be involved in the decision-making process, contributing to ethical guidelines that govern machine behavior. Users also bear the responsibility of understanding how these machines function and ensuring that their use adheres to ethical standards.

Organizations employing AI technologies must establish robust frameworks for ethical oversight. This practice includes audits of AI systems for fairness and compliance with ethical standards. At the organizational level, creating a culture of ethical awareness encourages responsible AI deployment and discourages harmful practices often motivated by profit.

Governments and regulatory bodies play a crucial role in legislating ethical practices in AI. Establishing frameworks for accountability, transparency, and fairness is essential. Regulatory bodies should monitor AI deployment and impose guidelines that ensure ethical considerations are integrated into technological development.

The role of society as a collective bears responsibility in advocating for ethical machines. Public engagement, awareness, and dialogue are vital for promoting a collective understanding of ethical implications in technology. Societal norms and values shape the trajectory of ethical considerations in machine development and deployment.

All things equal, there are *challenges* to these recommendations. The pace of rapid technological change often outstrips legislative and ethical standards. Policymakers struggle to keep up with innovations, leading to a regulatory lag that can make ethically questionable technologies permissible. The creation of AI government roles may help assuming that the persons are always current in the technologies and the impact on their citizens. Another challenge is conflicting ethical frameworks that complicate the ethical programming of machines. For instance, what happens when a machine confronts a situation involving a moral dilemma, such as prioritizing the life of a passenger over that of a pedestrian? Given different cultural perspectives on ethics, consensus on the "right" choice may be difficult to achieve,

An area not to be ignored is *data privacy and security*. Ethical considerations also extend to data privacy and security in AI systems. The extraction and utilization of personal data without consent may raise moral concerns about user privacy and autonomy. Establishing ethical norms in data management is crucial for ethical machine functionality.

7.3 Conclusion

In conclusion, the advent of ethical machines presents profound opportunities and challenges. Their significance lies in ensuring safety, fostering trust, and promoting fairness across societal structures. Responsibility is multilateral, encompassing developers, users, organizations, governments, and society, highlighting the need for collaborative efforts for responsible AI operation. To harness the full potential of ethical machines, stakeholders must actively engage in ethical deliberations, ensuring accountability and transparency. Thus, the future of technology should not be merely a pursuit of progress but an "Ethical Quest" prioritizing humanity's best interests in an AI-driven world.

References

[1] https://openai.com/blog/chatgpt/
[2] https://techcrunch.com/2022/12/30/theres-now-an-open-source-alternative-to-chatgpt-but-good-luck-running-it/
[3] https://www.npr.org/2023/03/31/1167491843/chatgpt-italy-ban-openai-data-collection-ai
[4] https://www.bbc.com/news/world-us-canada-65452940
[5] https://apnews.com/article/chatgpt-openai-ceo-sam-altman-congress-73ff96c6571f38ad5fd68b3072722790
[6] https://www.europarl.europa.eu/doceo/document/TA-9-2023-0236_EN.pdf
[7] https://www.wiley.com/en-us/What+To+Do+When+Machines+Do+Everything:+How+to+Get+Ahead+in+a+World+of+AI,+Algorithms,+Bots,+and+Big+Data-p-9781119278665
[8] https://brianchristian.org/algorithms-to-live-by/
[9] https://fastcompanyme.com/co-design/heres-why-human-centered-design-is-so-important/
[10] Machine Ethics, Anderson, Michael, Anderson, Susan Leigh Editors, Cambridge University Press, 2011.
[11] Ethical Machines, Your Guide to Totally Unbiased, Transparent and Respectful Aim Blackman, Reid, Harvard Business Review Press, 2022.
[12] https://www.moralmachine.net7

PART IV

Evoking the Humanized Internet Use Cases

DOI: 10.1201/9781003339724-12

8

Impact of Technology on Healthcare Today: A Nursing Perspective

Frances Hughes

Senior Research Fellow CHOPR Upenn, USA; Adj Professor QUT- Faculty of Health, School of Nursing, Australia

Today is an era of uncertainty and rapid change in healthcare. Across the globe, clinicians, hospital administrators, and funders are grappling with the increased demand and cost of services. At the same time, there are growing disparities within certain consumer groups and skill shortages. It can seem like a perfect storm – globalization, deregulation, privatization, health-care restructuring, medical tourism, and nursing shortages have all contributed to an increase in the focus on health-care systems and processes which serve to promote and validate the quality of nursing and health-care services. Technologies mentioned in this chapter on the whole refer to communication technologies.

For decades, the world of nursing and healthcare remained relatively isolated from the digital and technology revolution; but over the last several years, the world has become progressively smaller due to the increased interconnectivity gained by technological advances and other global risks. The interconnectedness of these risks is shown in Figure 8.1.

The Global Risk Report (2018) [1], developed from the Global Risks Perception Survey (GRPS), highlighted four concerns: persistent inequality and unfairness, domestic and international political tensions, environmental dangers, and cyber vulnerabilities in relation to healthcare; the risk connections mentioned in Figure 8.1 all affect us, and they underpin what we know as social determinants of health. The discussion on cyber vulnerabilities is particularly relevant to this writing. Although in previous years respondents to the GRPS have tended to be optimistic about technological risks, this year

DOI: 10.1201/9781003339724-13

Figure 8.1 Global Risk Report 2018 (World Economic Forum).

concerns jumped, and cyber attacks and massive data fraud both appear in the list of the top five global risks by perceived likelihood. Report outlines the types, prevalence, and disruption attacks are causing: "cyber breaches recorded by business have almost doubled in five years, and dark net markets for malware are on the resurgence."

Haefner (2017) [2] identified, through interviews with five health-care executives from organizations across the United States, that two of the top four challenges they faced were related to technology, namely the impact of healthcare's transformation to consumerism and the challenge of negative technology, cybersecurity, and confidentiality.

The first challenge is the ability to keep pace with healthcare's transformation to consumerism. Consumers now expect a similar customer experience from health-care organizations as from other business-to-customer interactions in their daily lives, like with hotels, airlines, and restaurants. This

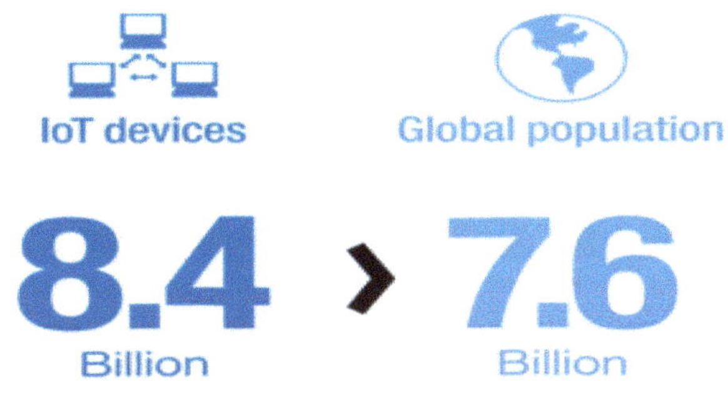

Figure 8.2 Global Risk Report.

shift in consumer attitude is one of the most pervasive changes in healthcare and therefore one of the top areas of concern. Processes that assist health-care managers to understand their consumers and demands have risen; patient engagement and data analytic tools used to understand cost estimates and bed utilization are now part of the hospital operational systems.The second challenge health-care executives face relates to navigating technology, cyber-security, and confidentiality:

> Technology is a double-edged sword... healthcare is regularly enamored by shiny new toys and buzzwords, like cloud technology, predictive analytics, AI, and blockchain. To keep focused and make the wisest investments, ... it is imperative that hospitals map out why they are investing in a certain solution, what the technology will solve and why previous solutions didn't work [2].

With the increased expectations of consumers to protect their data and protection from cyber attacks, many executives now spend more time identifying risks from technology to their business, and thus, their large technology investments have created risks in them by themselves. Increased interconnectedness, built on the advancement of technology, has caused a flood of investments; some have been stimulated by a country's desire to regulate and embrace these modern technologies. In the United States, the Health Information and Technology for Economic and Clinical Health (HITECH) Act 2009 stimulated growth in healthcare by 27 billion dollars [3], [4]. There is no doubt that technology has become a disruptor, an innovation that has an impact on the way we deliver healthcare.

Figure 8.3 2015 UN 17 Sustainable Development Goals.

For nurses, we have always had a wider world view on the determinants of health and the interconnectedness. This is reinforced in the 2015 UN 17 Sustainable Development Goals, which articulate the interconnectedness of health to wider social and economic systems [5].

The Sustainable Development Goals (SDGs), otherwise known as the Global Goals, are a universal call to action to end poverty, protect the planet, and ensure that all people enjoy peace and prosperity.

These 17 goals build on the successes of the Millennium Development Goals, while including new areas such as climate change, economic inequality, innovation, sustainable consumption, peace, and justice, among other priorities. The goals are interconnected – often, the key to success on one will involve tackling issues more commonly associated with another.

Technologies can promote quality if they are used appropriately and correctly. The ethics of application of many technologies are still debated in clinical groups, as to the role that the patients have in decisions about what are used in their treatments.

As a registered nurse of over 30 years, I have faced many challenges in my education and practice environment. The largest challenge has been keeping pace with the changes and speed of knowledge transformation. The rate and process by which information and new evidence travel around the world have meant as a nurse I had to learn to be more responsive and discerning. With the advancement of mobile phones and other hand-held communication devices, information that was once delivered at a snail's pace is now not days but minutes and seconds from informing and connecting clinicians and consumers. Internet and social media have played a significant role in informing the consumer but with this has come information overload and concerns regarding what information is reliable and accurate. Who do we, as practitioners, believe and how are things be authenticated, verified as "truth" and real evidence.

Advancements in technology while transforming the way we as nurse provide care does not come without challenges. From smart bedside solutions to providing education and remote monitoring of patients, it has never been more important for nurses and clinicians to keep pace with the rapid changes in health-care technology – designed to boost accuracy and efficiency and, ultimately, improve patient outcomes. Technology today should differ in one key area than the past; it should aim to put the patient/consumer in charge, no longer as passive recipients or bystanders of care; they need to be active agents.

The following section outlines challenges and opportunities of new technologies from a health perspective.

8.1 Challenges and Opportunities

Engelhardt (2018) [6] describes four fundamental challenges in health data and technology management: putting the patient at the center, privacy and access, completeness of information, and cost. Patients of today have greater expectations from providers regarding their care and security of their information, they also are costing more to provide the specialist care in our hospitals, and we operated in a system where data and information are predominantly siloed.

With personalized technological advancements, nurses today are faced with not only understanding the technology, applying the new diagnostics and treatments, but also ensuring that the care is patient centered, the voice of the patient is not lost in the technology and multidivisional nature of care provision. For the nurse, this is aggravated by the "time factor," how to maintain the level of skilled safe nursing to meet the patient's needs when there are staff shortages, constant reporting, and data collection by hospital administration. Technology combined with innovation has led to the explosion of health administrative systems and self-service devices all of which nurses are confronted with by either the hospital management or patients themselves [7].

8.2 Digital Health Technology

Digital health describes using digital information, data, and communication technologies to collect, share, and analyze health information for purposes of improving patient health and health-care delivery [8].

Electronic health record (EHR) access to patient's information is now right at the fingertips of nursing and medical staff, but with this has come a steep learning curve in relation to IT and data skills. The systemized collection of patient information in a digital format can be shared across many different health-care settings; they may include a range of patient data, including medical history, demographics, medication histories, allergies, immunization status, laboratory test results, vital signs recorded over time, radiology images, extensive personal information, and billing information and history. These EHR systems are designed to store data with greater accuracy and to capture and record the medical state of a patient across time and place of care [9].

The two barriers to address with EHR's are increasing interoperability (the ability to communicate, exchange data, and use the exchanged data) and patient centeredness [10]. Currently, systems are bespoke, don't talk to each other, and often siloed. Additionally, we have payment and provider systems which are often valued based on volume, not based on care; thus, the patient has little ability to access, contribute to their own records (wearable devices), have patient recorded outcomes, thus not deemed to be "patient centered." For clinicians, this is difficult to comprehend when their clinical practice protocols and guidelines and practice models are evidenced outcome based in many cases.

Secure and scalable data sharing is essential to provide effective collaborative treatment and care decisions for patients. No longer does a clinic have to track down a patient's previous paper medical records, as data entered into an EHR system is legible, therefore reducing the risk of error. These systems can also reduce the risk of data replication, because there is only one modifiable file. Therefore, the file is more likely to be up to date, and the risk of lost paperwork is also reduced. In addition, digital data are searchable, so clinical staff can extract medical data for the examination of possible trends and changes over time in a patient's health status. Population-based studies are also facilitated by the widespread adoption of EHR systems. In a perfect world, these digital health records may be used for the enhancement and benefit of everyone's health [11], [12].

Diagnostic tools: With the rapid growth in technology has come the development of machine-learning algorithms, similar to that used in Meta, Amazon, and Google (to optimize search queries). Self-service technologies such as apps and watches can now identify things such as heart irregularities [7].

Disease management and decision support tools: Sharma et al. (2018) [8] describe how these tools "can potentially improve patient outcomes by increasing patient engagement in self-care and care giver care, close communication gaps, and personalizing services to meeting patient needs." Applications are currently available for the management of diabetes insulin administration; the app requires a prescription, and then, the patient titrates the dosage through the use of the app [13].

Research: Technology has assisted greatly in the ability to connect to potential research participants. One example is MyHeart Counts Cardiovascular Health Study, which recruited 48,986 patients from March 2015 to October 2015 using Apple Research kit [14]. Informed consent is a critical part of research, apps are now available, and video consent process is proving popular [15].

Table 9.1 from [8] describes the various uses of digital health technologies in clinical research, from fundraising to recruitment and informed consent processes, as you can see the process is much more interactive and provides opportunities for more responsiveness and engagement with participants. The challenge I believe is in the ethical approval and the issue of storage post studies of information, which is a requirement of ethical approval. The concept of older data and information being held secure on USB or disks in locked filing cabinets is a thing of the past. How does one prevent data and information collected for one purpose being used for another?

Quality and safety: The challenge for today's clinicians is to manage time and expectations and maintain quality and safety. With previous paper-based systems, the task of report writing, reviewing, and actioning occurred across days and weeks, but now it occurs in minutes and hours. Nurses frequently report on the conflict between providing care to the patient and ensuring documentation is complete. Most doctors, nurses, and clinicians can attest to a significant increase in their cognitive workload during EHR system implementation, resulting in a definite decrease in productivity. Clinicians also continue to note their concerns over safety being compromised, either through the speed by which information they are expected to respond to data or the systems that generate it [16].

Safety and quality are key concerns. Vincent (2010) [17] defines patient safety as the "avoidance, prevention and amelioration of adverse outcomes

Table 8.1 Potential uses of digital health technologies in clinical research.

Stage of trial	Potential use (Examples)	Successful use cases
Fundraising	Crowd funding through KickStarter or GoFundMe	Oncolytic virus treatment for neuroendocrine tumors raised over $3 million through crowdfunding
Recruitment	Use of apps and social media (e.g., X/Meta).	MyHeart Counts Cardiovascular Health Study enrolled 48,968 subjects in 6 months
Consent and patient surveys	Mobile devices enable rapid data capture	The PALM registry used mobile devices to obtain patient consent and conduct surveys
Obtaining novel endpoints	Digital technologies can obtain information that might detect more benefit or harm than conventional trial endpoints	NEAT-HFpEF trial identified potential worsening of physical activity with nitrate use by using accelerometers. This adverse outcome was not identified through the 6-min walk test
Remote patient monitoring	Cardiac rhythm monitoring	Most experience with implanted devices (e.g., ICDs in cardiac rhythm monitoring)
	Remote hemodynamic monitoring of physiological parameters might allow for assessment of drug titration and can guide therapies	The implantable pulmonary artery pressure sensor device (CardioMEMS) identified a reduction in heart failure hospitalization in the CHAMPION trial
Development of patient communities throughout the conduct of a clinical trial	Encourage retention, recruitment, and knowledge dissemination	The MyHeart Counts cardiovascular health study enables patients to rapidly join the study through the mobile app while contributing data and learning about cardiovascular disease prevention

CHAMPION = CardioMEMS Heart Sensor Allows Monitoring of Pressure to Improve Outcomes in NYHA Class III Heart Failure Patients; ICD = Internal Cardioverter-Defibrillator; NEAT-HFpEF = Nitrate's Effect on Activity Tolerance in Heart Failure with Preserved Ejection Fraction; PALM = Patient and Provider Assessment of Lipid Management.

Source: Sharma et al. (2018, p. 2684).

or injuries stemming from the process of healthcare." As health IT systems become the core parts of our hospital services, failures of these systems can lead to patient harm, and thus, the safety of systems is paramount [18]. The portability and accessibility of EHR systems may also have unintended consequences as data, as mentioned previously, can be accessed and stolen by unauthorized or criminal users more easily than the old-fashioned paper medical record, sitting locked in a file cabinet in providers office.

The 2011 Institute of Medicine report, Health IT and Patient Safety: Building Safer Systems for Better Care, found the literature about health IT and patient safety to be inconclusive; some deliver benefits to patients, that is, medication administration, while others create new hazards. Electronic prescribing can help eliminate error that occurs through eligible handwriting, but IT systems through shortcuts can create unintended risks [19]. IOM in their 2011 report strongly advocated that products do not exist in isolation but need to view them as part of larger sociotechnical systems, and thus, safety emerges from the interactions of these factors. They concluded their report by stating:

> To achieve better healthcare, a robust infrastructure that supports learning and improving the safety of health it is essential [16],

The American Medical Informatics Association (AMIA) made recommendations in their 2010 Board Position Paper to address these concerns. Their recommendations include ethics education for buyers and users of EHR systems, transparency guidelines, development of and adoption of best practices, and examination of current regulations dealing with electronic health records and applications. Issues like conflict of interest and certainly patient privacy concerns were raised in light of the way the doctor–patient relationship is affected by an electronic intermediary [20].

Perhaps most importantly, privacy and security concerns have slowed the adoption of EHR. A review of the national HIT safety initiates concluded that safety programs "are at varying different stages of maturity, with England having the longest standing and most well-developed safety programs while Canada and USA are at earlier stages" [21].

In the European Union, new regulations came into place in 2018 to protect the processing of personal data, including health-care data. Threats to health-care information can be generally categorized as:

- Human threats, such as rogue employees or inappropriate access and use of information
- Environmental threats, such as earthquakes, fires, and hurricanes
- Technology failures, such as a system crash.

All these threats to health-care data security can be internal or external, intended or unintended, and health-care professionals will always have these threats in mind when discussing ways to protect their patients' private information. The Health Insurance Portability and Accountability Act (HIPAA) has attempted to mitigate these security threats by developing a framework that is comprehensive but not so specific that it limits the options of clinicians who may have access to alternative technologies. In Canada, the Personal Information Protection and Electronic Documents Act (PIPEDA) of 2000 established rules on the use, collection, and disclosure of personal information. Information affected includes both non-digital and electronic records.

There is a concern with the safety risks: EHR will result in increased compliance costs as providers and insurers protect themselves from legal and malpractice claims. It is thus important for clinicians to have more education on the systems being implemented and also time to ensure that do provide quality documentation and data into them.

Regardless of the challenges of interoperability, patient centeredness, and security, the current data sharing systems require patients to obtain and share their own medical records with providers. This is a complex system requiring consent, often slow, incomplete (due to siloed systems), and insecure. New approaches are needed, one of which is blockchain.

8.3 Blockchain

> Blockchain is a way for people to solve problems by sharing things
> Diego Espinosa Founder and CEO Healthcoin [6].

As previously stated, healthcare today suffers from siloed and fragmented data, delayed communications, and disparate workflow tools caused by lack of interoperability. Blockchain offers the ability to enable access to longitudinal, complete, and tamper-aware health records that are stored in fragmented systems in a secure and pseudo-anonymous fashion [22].

Blockchain is now classed as the largest technological and business revolution since the advent of the internet, and economists view it as the 4th revolution of history. Its growth has come from the need for heightened security, improved fraud prevention, transactions management, and a more direct method for individuals to manage their own data. It is broadly understood to be on the same scale as the internet and artificial intelligence in its ability to impact the future. The ability to connect stakeholders to data without an intermediary and its associated costs is attractive.

Simply put, blockchain is a technology that allows for the secure storage, communication, and exchange of data. However, unlike today's databases, it has some key differences:

- It is distributed allowing participants to see and verify all of the data involved. This minimizes chances of fraud because authorized participants can verify all the data and transactions contained in any one block of the chain.
- All data are verified and encrypted before being added to the block. Again, this process is managed publicly and organized in such a way that a hacker would need to have exponentially more computing power to break in than is required in today's systems.
- These two characteristics allow blockchain to provide public ledgers that participants trust based on the agreed-upon consensus protocol described above. Once data (transactions) are on a block, they cannot be overwritten, and no central institution holds control, as these ledgers are visible to all the users connected to them. Users' identities within a ledger are known only to the users themselves.

Strategically, blockchain will reduce the cost of trust. Individuals will be able to interact more directly with each other without the participation of institutions, that is, research communities can be established:

> It offers a way for people who do not know or trust each other to create a record of who owns what that will compel the assent of everyone concerned. It is a way of making and preserving trust, it has far reaching implications beyond financial. It is the blockchain that replaces trusted third parties [23].

Zhang et al. (2018) [22] identified seven areas in which blockchain has uses in healthcare:

1. Prescription tracking to detect opioid overdose and over prescription.
2. Data sharing to incorporated telemedicine with traditional care.
3. Sharing cancer data with providers using patient authorized access.
4. Cancer registry sharing to aggregate observations in cancer cases.
5. Patient digital identity management for better patient record matching.
6. Personal health records for accessing and controlling complete health history.

7. Health insurance claim adjudication automation to surface error and fraud.

In the future, we will not only see blockchain technology being used in relation to hospital services and patients but also see it being used in our profession. Universities are now awarding their qualifications on to blockchain technology; thus, graduates can then determine who has access to their verified and authenticate award (University of Nicosia). There will be no need of third-party agencies like notaries to verify. This will extend also in future to continuing education credits as providers embrace this technology. No more issuing of paper certificates which are not verified or authenticated. Our nursing license in the future will also be awarded in this manner, and for nurses working between states, countries will to send their blockchain link to regulatory and credentialing bodies or direct to employers without intermediaries. Hospital administrators spend hours with regulators and professional colleagues, establishing credentials of clinicians. This will be a major cost saving and quality improvement for this process [24], [25]. The potential disruption this will cause to organizations that rely on antiquated systems reliant on paper-based processes and manually driven will be immense. Nursing regulators, credentialing organizations, and professional bodies be warned, disruption has stared and will be coming your way soon.

Engelhardt (2018) [6] describes how IT requires careful technical and administrative planning that includes all stakeholder. The critics of blockchain talk about the issue of security and how do you know it is secure? The response is that there is no guarantee of future performance but the fact that this technology has been around with Bitcoin open to hackers with a lot to gain but remains secure is important to note.

For nurses and other health professionals, we need to advocate for the patient to be in control of their data, but with that we need to ensure they have the ability to manage this through access to education on this new technology.

8.4 Artificial Intelligence

Artificial intelligence (AI) has evolved since it was founded as an academic discipline in 1950s. As machines have become increasingly sophisticated, tasks that are used to require "intelligence" are removed from the definition of AI. For example, optical character recognition has become a routine technology. Some machine capabilities generally classified as AI include successfully understanding human speech, competing at the highest levels in games like chess, self-driving cars, and intelligent routing systems used in delivery networks and military operations. For most of its history, AI has been pursued via sub-fields based on technical considerations, particular and specific goals, the use of specific tools, and even social factors, such as the particular interest of researchers [26], [27].

AI researchers have been able to create many tools that can perform jobs that are difficult and time-consuming for people to do, but they have also struggled to create a computer capable of carrying out tasks that are simple for people to do. For instance, a computer can easily and quickly perform complex mathematical calculations, but so far is unable to complete a complex human task like making a cup of coffee in a stranger's kitchen. The failed predictions of AI researchers and a lack of a complete understanding of human behaviors have diminished the original idea of human-level AI, although some researchers still believe it is ultimately possible. The intricacy of scientific problems and the need to fully understand the human brain through neurophysiology and psychology have limited AI researchers from emulating the function of the human brain in a computer hardware and software [28], [29], [30].

In the health-care field, the following are some examples of how AI is currently being implemented in hospital and care settings:

Robotics: It is not uncommon in today's hospital settings to see robots. These are being used to analyze data, administer medication, and undertake surgical procedures. Heartland is one such robot; this is a miniature robot that enters through a small incision on the chest to perform mapping and therapy over the surface of the heart. The impact is said to be to the value of $40 billion dollars to healthcare [31], [32].

Virtual nursing assistants: From interacting with patients, providing information on key health topic to directing patients to the most effective care setting, virtual nurses are available 24/7; they can answer questions, monitor patients, and provide quick answers. Most applications of virtual nursing assistants today allow for more regular communication between patients and care providers between office visits to prevent hospital readmission or unnecessary hospital visits. Care Angel's virtual nurse assistant can even provide wellness checks through voice and AI [30].

Aid clinical judgment or diagnosis: While AI in hospital settings is in its infancy, there are some exciting developments. A Stanford University study tested an AI algorithm to detect skin cancers against dermatologists, and it performed at the level of the humans [33]. Bloomberg News (2018) [34] announced early this year that a Danish AI software company tested its deep-learning program by having a computer eavesdrop while human dispatchers took emergency calls. The algorithm analyzed what a person says, the tone of voice, and background noise and detected cardiac arrests with a 93% success rate compared to 73% for humans [28], [35].

Workflow and administrative tasks: Technology such as voice-to-text transcriptions could help order tests, prescribe medications, and write chart notes. One example of using AI to support admin tasks is a partnership between the Cleveland Clinic and IBM that uses IBM's Watson to mine big data and help physicians provide a personalized and more efficient treatment experience. Watson supports physicians to analyze thousands of medical papers using natural language processing to inform treatment plans [27], [35].

Image analysis: Currently, image analysis is very time-consuming for human providers, but an MIT-led research team developed a machine-learning algorithm that can analyze 3D scans up to 1,000 times faster than what is possible today. This near real-time assessment can provide critical input for surgeons who are operating. It is also hoped that AI can help to improve the next generation of radiology tools that don't rely on tissue samples. Additionally, AI image analysis could support remote areas that don't have easy access to health-care providers and even make telemedicine more effective as patients can use their camera phones to send in pictures of rashes, cuts, or bruises to determine what care is necessary [28], [27], [30], [35].

In the very complex world of healthcare, AI tools can support human providers to provide faster service, diagnose issues, and analyze data to identify trends or genetic information that would predispose someone to a particular disease. When saving minutes can mean saving lives, AI and machine learning can be transformative not only for healthcare but also for every single patient.

8.5 Standards and Ethics

While continued research on the use of AI in healthcare hopes to validate its effectiveness in improving patient outcomes, its use may also introduce several new risks to patients and health-care providers. These risks include algorithmic (and data) bias, Do Not Resuscitate implications, and other "machine morality" issues. These potential challenges to the use of AI have prompted the call for regulations on its use [36].

As of this writing, there are no regulations on the use of AI in healthcare. In May 2016, the White House announced its plan to host a series of workshops and the formation of the National Science and Technology Council (NSTC)

Subcommittee on Machine Learning and Artificial Intelligence. The group published The National Artificial Intelligence Research and Development Strategic Plan in October 2016, which outlined its proposed priorities for federally funded AI research and development. This R&D is within government and academia at this point. The Federal Drug Administration (FDA) has expressed concern. Bakul Patel, the Associate Center Director for Digital Health of the FDA, has quoted saying in May 2017:

> We're trying to get people who have hands-on development experience with a product's full life cycle. We already have some scientists who know artificial intelligence and machine learning, but we want complementary people who can look forward and see how this technology will evolve [37], [30].

Given the many medical data breaches recently reported, and the lack of public trust in the ultimate security of centralized systems, some countries have enacted laws requiring safeguards to be put in place to protect the security and confidentiality of medical information as it is shared electronically and to give patients some important rights to monitor their own medical records and receive notification for loss and unauthorized acquisition of health information. The United States and the European Union have imposed mandatory medical data breach notifications. While a patient's medical information can be shared by a number of people both within the health-care industry and beyond, the Health Insurance Portability and Accessibility Act (HIPAA) of 2003 [38] is an U.S. Federal Law pertaining to medical privacy [39].

It established standards for patient privacy in all 50 states, including the right of patients to access their own records. It is bound by the Code of Ethics of the medical profession and also by the legislation on data protection and criminal law. The maintenance of confidentiality and privacy of patients applies first to the patient's medical history, which must be adequately safeguarded, accessible only to the authorized personnel. However, the precepts of privacy must be observed in all fields of hospital life: privacy at the time of the conduct of the anamnesis and physical exploration, the information provided to relatives, the conversations between health-care providers in the corridors, maintenance of adequate patient data collection in hospital nursing controls, telephone conversations, open intercoms, and so on. While HIPAA provides some protections, it is pertinent to note that between 2006 and 2012, medical and health-care providers experienced 767 security breaches resulting in the compromising of confidential health information for 23,625,933 patients [39].

As digital technologies become less expensive and readily available to the public directly, greater oversight from regulators will occur. Sharma et al. (2018) [8] suggest that it is timely for the establishment of a framework for their appropriate use to occur.

8.6 The Role of Patient

Patient-centered care is now an expected approach articulated from WHO through to professional practitioner organizations. As stated previously, there is increasing agreement that patients should be as active as possible as agents in their own care, not just mere passive recipients [40], [41]. When the procedures and targets of a health-care institution eclipse local concerns, patient neglect is possible. Various hospital investigations have shown the dangers of silencing the voices of patients, who are in the best position to notice problems such as service delays, poor hygiene, poor conduct, and poor communication.

Technology will play an increasing role in prevention and health promotion, the self-service technologies will be more widely accepted in our hospital settings, and patients will expect them to integrate into EHR. They also will expect to control their own data and who then can access it. The customer (patients) will need access to education to be able to discern and understand the technology they are accessing and also a health navigator to enable decision to be made out of the options this technology will provide.

The Global Risk Report (2018) [1] and the the IOM (2011) [16] report both highlight the information overload that is occurring not only for consumers relying on health-care options but also for practitioners. Studies have found that people have a hard time distinguishing between accurate and fake headlines. One survey in late 2016 presented respondents with a random selection of six headlines – three accurate and three false – and asked them to rate the accuracy of the headlines they could recall having seen before. It found that 75% of the time respondents judged the false headlines to be "somewhat accurate" or "very accurate" – only slightly lower than 83% for the accurate news headlines [1].

In relation to discussion on the application of blockchain, the following is a salient comment:

> If the whole point was to empower the patient and address the failures of our current system, which rely exclusively on the patients

trust, (and if we create a new system where patients are empowered to control their data but do not know what to do with it and end up engaging other to manage it for them), are we really moving away from a trust-based management system? (Tay, N., personal communication cited in [6]),

8.7 Conclusion

Nurses have been involved with informatics for decades [42]. The Healthcare Information and Management Systems Society defines nursing informatics as "a specialty that integrates nursing science, computer science, and information science to manage and communicate data, information, knowledge, and wisdom in nursing practice." The issue of today is the speed at which this technology is being adopted and the ability to directly interface with consumers and patients without the need for intermediary like health organizations and practitioners like nurses.

By the end of 2020, it is predicted that there will be over 50 billion connected devices, all of which set the stage for millions of data transactions. As we move forward in this connected, digital first world, data are becoming a new kind of currency, or more so, we are moving into a data economy. But in this new economy, privacy, which is a fundamental human right, becomes harder to maintain. With the Internet of Things and applications such as wearable personal health trackers, consumers are becoming the center of their care management. So, what then does this mean for us as nurses? Are we engaged to support our patients and consumers? Do we indeed understand the impact of this technology?

The role of the nurse as it has done for centuries will change as the patients' needs change; we will have to educate, adapt, and apply our new skills and knowledge to address these changes. Key to this will be the way nurses can be present and engage with patients and consumers, using technology as a tool, not as a replacement for humanized healthcare. There will be certain roles and functions that we as health professionals will no longer be required to undertake, and we will need to learn to let go and let consumers themselves undertake or agree for them to become obsolete. We will need to ensure that both our professional continuing education is kept up to date and also we advocate for access to education on these new technologies and their applications for our patients. The future of nursing is safe if we adapt and keep close to our patients; it is our depth and engagement with our patients that will ensure our future as well as their future.

References

[1] Global Risk Report. (2018). World Economic Forum retrieved from http://reports.weforum.org/global-risks-2018/global-risks-landscape-2018/

[2] Haefner, M. (2017). *R*etrieved from https://www.beckershos*pitalreview.com/hospital-management-administration/4-of-the-most-press*ing-challenges-health-systems-

[3] Aaron, H. *(2018). Heath care* costs and medical technology. Retrieved from https://www.thehastingscenter.org/briefingbook/health-care-costs-and-medical-technology/

[4] Rock Health. (2017). Rock Health Report. Retrieved from https://rock-health.com/report-category/report/accessed 19/11/2018

[5] United Nations. (2015). Sustai*nable Development Goals. Retrieved from https:*//sustainabledevelopment.un.org/?menu=1300

[6] Egelhardt, M.A. (2017). Hitching healthcare to the chain: An introduction to blockchain technology in the health care sector. Technolog*y Innovation Management Review, 7(10).*

[7] Immonen, M. and Koivuniemi, J. (2018). Self-service technologies in health care: Exploring drivers for adoption. Computers in Human Behaviour 88, 18–27.

[8] Sharma, A., Harrington, R.A., McClellan, M.B., Turakhia, M.P., Eapen, Z.J., Steinhubl, S., ... Peterson, E.D. (2018). Using digital health technology to better generate evidence and deliver evidence-based care. Journal of the American College of Cardiology, 71(23): 2680–2690. https://doi.org/10.1016/*j*.jacc.2018.03.523

[9] Showell, C.M. (2011). Citiz*ens, patients and policy: a challenge for Aus*tralia's national electronic health record. Health Information Management Journal, 40(2): 39–43. doi:10.1177/183335831104000206. P*MID 28683627. http://www.himaa.org.au/members/journal*

[10] Reid, R., Compton*, D., Grossman, J., Fanjiang, G. (2005)*. Building a better delivery system: A new engineering/health care partnership. National Academies Press, Washington.

[11] Ferguson, T. (1995). Consumer health informatics. Health Forum Journal 38:28–33.

[12] Medical info*rmatics. (2018). Retrieved 11/2/2018* from https://en.wikipedia.org/wiki/Health_informatics

[13] Bluestar. (2017). Retrieved from http://www.bluestardiabetes.com

[14] McConnell, M., Shcherbina, A., Pavlovic, A. (2017). Feasibility of obtaining measures of lifestyle from a smartphone app: The MyHeart Counts Cardiovascular Healt*h Study. JAMA Cardiol*, 2:67–76.

[15] Grady, C., Cummings, S., Rowbothan, M., McConnell, M., Ashley, E., Kang, G. (2017). Informed consent. New England Journal of Medicine; 376:56–*57.*

[16] Institute of Medicine. (2011). Health IT and patient safety: Building safer systems for better care. R*etriev*ed from https://essentialhospitals.org/wp-content/uploads/2014/07/IOM-report-on-EHR-and-Safety.pdf

[17] Vincent, C. (2010). Patient safety. 2nd edition. Wiley Blackwell. Oxford

[18] Habli, I., White, S., Sujan, M., Harrison, S., Ugarte, M. (2018). What is the safety for health IT? A study of assurance practices in England. Safety Science 110, 324–335.

[19] Bates, D., Leape, l., Cullen, D., Larid, N., Pet*ersen, L., Teich, J., Burdick, E., Hickey, M., Kleefield, S., Shea,* B., Vander, V. (1998). Effect of computerized physician order entry and a team intervention on prevention of serous medication error. JAMA, 280 (15): 1311–1316.

[20] American Medical Informatics Association. (2018). Retrieved from www.amia.org

[21] Kushniruk, A., Bates, D., Bainbridge, M., Househ, M., Boryycki, E. (2013). National efforts to improve health information *systems safety in Canada, the United Sta*tes of America and England. International Journal of Medical Information, 82(5): e149–e160.

[22] Zhang, P., Schmidt, D., White, J., Lenz, G. (2018*). Blockchain technology use cases in he*althcare. Retrieved from https://doi.org/10.1016/bs.adcom.2018.03.006

[23] The Economist (2015). The great chain of being sure about things. Retrieved from https://www.economist.com/briefing/2015/10/31/the-great-chain-of-being-su*re-about-things*

[24] Althauser, J. (2017). "Illinois Government *Pilots Blockchain T*echnology for Birth Certificate Digitization." CoinTelegraph: The Future of Money. Web.

[25] Smolenski, N. (2018). "Introducing Blockcerts." YouTube. YouTube, LLC, 22 February 2018.

[26] Deloitte. *(2018) https://www2.deloitte.co*m/content/dam/Deloitte/de/Documents/Innovation/Artificial-Intelligence-Innovation-Report-2018-Deloitte.pdf accessed 13/11/2018

[27] Kalis, B., Collier, M., Fu, R. *(2018). 10 p*romising AI applications in healthcare. Harvard Business Review. https://hbr.org/2018/05/10-promising-ai-applications-in-health-care

[28] Begg, R. (2009). Artificial intelligence techniques in medicine and healthcare. In J. Tan (Ed.), Medical informatics: Concepts, methodologies, tools and applications. IGI Global.

[29] De Sumoni, A., Taylor, S., D Phil, C., Panzarasa, P., Sheikh, A. (2018). Online "superusers" as allies of the health care workforce. Retrieved from https://catalyst.nejm.org/superusers-allies-online-heal*th-communities/*

[30] Makridakis, S. (2017). The forthcoming artificial intelligence (AI) revolution: Its impact on society and firms. Futures, 90(Supplement C), 46–60. https://doi.*org/10.1016/j.futures.*2017.03.006.

[31] Dirican, C. (2015). The impacts of robotics, artificial intelligence on business and economics. Procedia - Social and Behavioral Sciences, 195, 564–573.

[32] VentureBeat. (2018). The 3 most valuable applications of AI in health care. Retrieved from https://ve*nturebeat.com/2018/04/22/the*-3-most-valuable-applications-of-ai-in-health-care/

[33] Stanford News. (2017). Retrieved from https://news.stanford.edu/2017/01/25/artificial-intelligence-used-identify-skin-cancer/

[34] Bloomberg News (20*18). Retrieved from https://www.bloomberg.com/* news/articles/2018-06-20/the-ai-that-spots-heart-attacks

[35] Plumb, J., Hains, I., Parr, M., Milliss, D., Herkes, R., Westbrook, J. (20*17). Te*chnology meets tradition: The perceived impact of the introduction of information and communication technology on ward round in the intensive care unit. International Journal of Medical Informatics, 105, 49–58.

[36] Verghese, A., Shah, N., Harrington, R. (2018*). What this computer n*eeds is a physician: Humanism and artificial intelligence.

[37] Artificial Intelligence. (2018). Retrieved from https://en.wikipedia.org/wiki/Artificial_intelligence

[38] HIPPA. Retrieved from www.hhs.gov/hipaa/for-professionals/privacy/laws-regulations/index.html

[39] EMR records and medical privacy. (2018). Retrieved on 11/3/2018 from https://en.wikipedia.org/wiki/Medical_privacy#Electronic_Medical_Records_(EMR)-fortune20160515-2

[40] Kitson, A., Marshall, A.*, Basssett,* K., Zeitz, K. (2013). What are the core elements of patient-centred care? A narrative review and synthesis of the literature from health policy, medicine and nursing. Journal of Advanced Nursing, 69: 4–15.

[41] Stewart, M. (2001). Towar*ds a global definition of patient centred care: The patient should be the jud*ge of patient centred care. BMJ, 322(7284): 444–445.

[42] Bakken, S. (2001). An informatics inf*rastructure is essential for evidence-based p*ractice. J Am Med Inform Assoc, 8(3):199–201.

[43] Blockchain. (2018). Retrieved from https://en.wikipedia.org/wiki/Blockchain#cite_note

[44] Healthcare Information and Management Systems Society. Nursing informatics 101. http://www.himss.org/ fi les/HIMSSorg/handouts/NI101.pdf

[45] Misener, J.H. (1990). The *impact of technology on the qu*ality of health care. Quality Review Bulletin, 16 (6): 209–213.

9

eGovernment as a Service

Daniel Gasteiger and Adithya Pradeep Kumar

9.1 Introduction

The liberal order which shaped society in the aftermath of the World War II and delivered peace and prosperity on an unprecedented scale, today faces a crisis. A crisis of underrepresentation. In the years following the financial crisis, public sentiment has rightfully turned against the current order. We believe that the crisis stems from a hijack of the current order by the elites and an underrepresentation of the needs of the masses in government. Today's technology presents an opportunity to change that. With the digital tools at our disposal today, we envision a future where the voice of every citizen in the world will be heard and we hope to contribute to that future by building the technological foundations of tomorrow's government.

9.2 The Crisis of Liberalism – Hijack by the Elites

Liberalism, a moral and political philosophy which inspired the great revolutions of the 18th century – the French Revolution and the American Revolution – has played a pivotal role in shaping modern society. A philosophy which finds its formal roots in the writings of John Locke, liberalism has gone through a mercurial history before emerging as a dominant architect of society following World War II and further consolidating its position with the fall of the Soviet Union [1], [2].

Through its long history, liberalism has been construed in a multiplicity of fashions. In fact, its longevity could be attributed to its constant reinvention (Bell, 2014 p. 22). In this context however, liberalism is interpreted as a philosophy which espouses equality, civil liberties, democracy and free markets backed by a welfare-state and policy interventions to counter market-failures. Since the early 1980s, the number of countries governed by democratic regimes has more than doubled [3]. The decade following the 1980s also saw

DOI: 10.1201/9781003339724-14

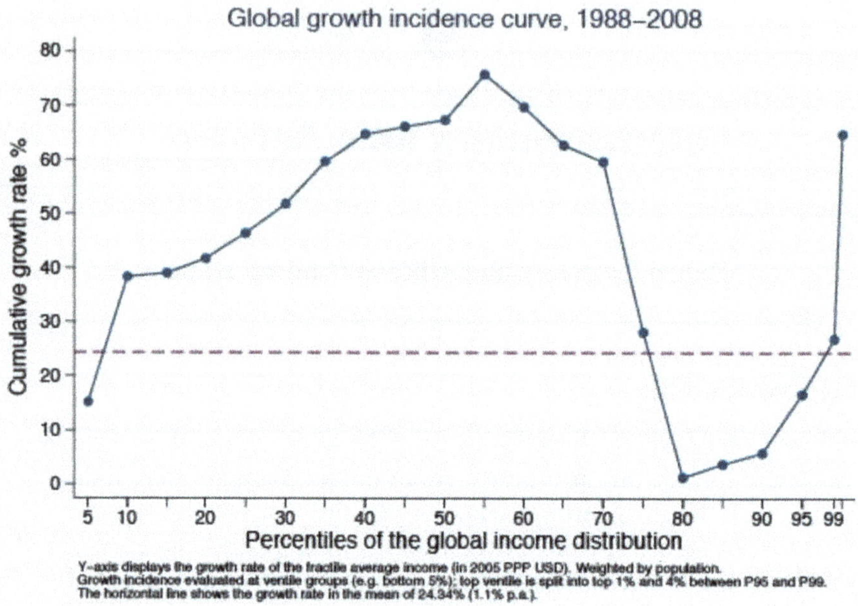

Figure 9.1 Anonymous global growth incidence curve: Real income change at various percentiles of the global income distribution between 1988 and 2008 (%). Source: Lakner and Milanovic (2014).

a widespread global move toward open markets and privatization of state-owned enterprises in countries ranging from the United Kingdom to India, which brought with it strong economic growth. Furthermore, since 1981, the proportion of people living in poverty across the world has fallen from 42.3% to 10.9% in 2013 [4].

However, while liberal ideologies have driven prosperity and civil liberties previously unknown to mankind, it is hard to ignore the many people who were left behind. The graph below tells the tale of both the success and the failure of the global economy in the aftermath of the fall of the Berlin Wall, leading up to 2008. Over this period, the greatest growth in real income is seen around the median, indicating the emergence of a strong middle class. At the same time, one also sees that growth in real income among the most vulnerable populations (the bottom 5%) falls below the mean growth across the global population. Even more striking is the staggering growth among the top 5% which accounted for 44% of the growth in real global income over the period [5].

While one would expect that the global financial crisis of 2007-08 which toppled behemoths such as Bear Stearns and Lehman Brothers would have had a levelling impact on global inequality, more recent figures tell a different story. As of 2010, 388 individuals owned as much wealth as the bottom 3.6 billion people in the world. If that figure is alarming, in the five years that fol- lowed, the 62 most wealthy individuals saw their wealth rise by 45% and by 2015, they accounted for as much wealth as the bottom 3.6 billion people [6].

Liberalism, a philosophy which found its roots in disrupting established power centres and is marked by a strong distrust of concentration of power, has now spawned new power centres. As of December 2024, the five most valuable companies, all tech-giants, boasted a combined market cap of over 10trillion USD, nearly thricethat of India's GDP – a country of 1.3 billion people.

Each of these companies in their own ways have driven democratisation of information on a scale never seen before. But the sheer scale of these organizations coupled with their questionable tax avoidance practices and the negligent management of personal data of billions of users, raises questions about their impact on society. A European Commission report from 2017 stated that the effective tax rate of multinational tech-businesses in the EU28 stood around 10% compared to 23.2% for traditional businesses. Social media platforms have been used by miscre- ants to incite violence, including against the much-persecuted Rohingyas in Myanmar to disrupting the handover of presidential power in the United States. The Cambridge Analytica scandal also brought to light the possibility of exploiting these platforms to distort public opinion and consequently democratic processes. Research indicates that two-thirds of American adults received their news in 2017 through social media, making it the most suitable medium for propagandists to shape public opinion. Further, the tsunami of free and often fake information provided by digital media, is threatening the sustenance of traditional media outlets which have historically been a vital pillar of democracy.

The series of events starting from the financial crisis of 2007-08 from which the elites emerged largely unscathed, and the economic stagnation and growing inequality which ensued has rightfully turned public sentiment against the current order. Couple the large-scale discontent, with social media platforms which permit actors from across the political spectrum to utilize user-generated data to guilefully target unassuming citizens with personalized and triggering content, and you have the perfect recipe for extreme polarisation in public opinion. Populists from the extremes of the political spectra capture the imaginations of people while the liberal elites stare indignantly at the phenomena, failing to recognise the fact that the current order is out of touch with the needs and voices of the masses.[8]

This introduction seeks not to serve as a critique of liberalism, populism or the tech giants – that's best left to the experts – rather, it seeks to bring to light the need to reinvent government to make it more responsive and adequately representative of the people who elect them. The problems on display today are reflective of the fact that the masses have been left out of the dialogues which inform policy making and shape society, skewing the current order in favour of the elites.

The technological tools at our disposal today allow us an opportunity to redefine government while placing the citizen's interests at the core of the design process. A more digital and citizen centric government is a fundamental requirement for a Humanized Internet. Building a Humanized Internet is not a choice but a moral imperative. This chapter outlines a future for digital government which put power firmly back into the hands of the citizens.

9.3 Digital Identity – The Foundation for a Digital Society

To know what technology can do to transform governance and empower people, one need look no further than the tiny Baltic nation of Estonia. Emerging independent from Soviet rule in 1991, Estonia set out to modernise its government and society, placing the still nascent internet at the core of its strategy. Today, nearly three decades later, Estonia is arguably the most digital society in the world. With 99% of public services accessible online, the country saves 800 years' worth of working time every year [9].

The most fundamental building block of this digital society is the national e-Identity. With 1.307 million active e-Identities (as of September 9, 2018), the country boasts near universal access to digital identity for its citizens. These e-Identities are at the core of trusted interactions between citizens, governments, and businesses, with over 24 million transactions every month.

The example of Estonia above displays the importance of e-identity in building a truly digital society where citizens benefit immensely from the increased efficiency and accountability in government. Elsewhere in the world, India is approaching the end of the first decade following the implementation of Aadhaar, the country's own e-Identity program. The impact there has been no less remarkable.

With 1.22 billion unique Aadhaar numbers issued to date, for the first time in the country's history, well over 90% of the country's residents now possess a legal identity [10]. This has supported financial inclusion on an unprecedented scale in the country with 325 million new bank accounts opened in the last four years [11]. The explosion in access to bank accounts has also enabled the government to launch a Direct Benefit Transfer (DBT) program, through which subsidies are deposited directly into the bank accounts of beneficiaries, thereby reducing leakages in subsidies through fake identities.

While Estonia and India are among the more advanced countries in terms of deployment of national digital identity programs, a number of other suc- cess stories are beginning to emerge across the world. But these centralized identity programs are not without challenges. Both Estonia and India have seen their e-Identity programs plagued by security and privacy concerns,.

9.4 The Path to Universal Digital Identity & Inclusion

Recent technological developments present an opportunity to reimagine identity in the digital realm. The proliferation of smartphones and the internet, aided by blockchain technology, offers the opportunity to put identity in the digital realm firmly back in the hands of the citizens. With smartphones, citizens can now have their government attested identity data stored securely on their own devices with a mathematical proof of the validity of this data stored on the blockchain.

The implications of this are potentially revolutionary. Businesses and service providers can now digitally verifythe identities of citizens without having to rely on a centralized database which guarantees the legitimacy of this data. This offers citizens streamlined access to a range of services in the digital & physical realm without the risk of being tracked across services and being comprehensively profiled, as is the in today's world of centralised identities.

While blockchain holds immense potential to be a privacy-enhancing technology, one must couple it with key technical design considerations so as to uphold the self-sovereignty of a citizen's digital identity. In addition to limiting the centralized storage of data to the minimum necessary, citizens must be empowered with explicit consent tools while sharing their data. In line with values of the blockchain movement, public IT infrastructure such as digital identities, have a lot to benefit from an open-source approach. Beyond the ethical imperative of transparency, open source also opens up the code base to public scrutiny, allowing it to evolve continually and grow in robustness.

But when one steps back to look at the state of identity across the world, it is alarming to learn that there are 1.1 billion people globally who lack legal identity. That's a seventh of humanity living an invisible existence, struggling to access the most basic services that many of us take for granted.

A deeper evaluation of the cause for this exclusion, prompts a realization that the ideal of universal access to identity will remain elusive if we rely on governments as the sole issuers of identity. As illustrated by the Peruvian example where the cost of enrollment per citizen can rise as high as USD 80 in the country's remote corners, the financial cost of reaching every last citizen with identity can be prohibitively expensive for governments [13]. Still worse, certain governments actively pursue discriminatory policies against select ethnic groups, thus excluding them from their identity programs and society at large. The case of Myanmar's ethnic minority, the Rohingyas is a case in point.

At the same time, the world is ripe with examples of NGOs, social enterprises, and supranational organizations among others reaching citizens where governments fail to. For instance, 86% of infants across the world receive routine immunization whereas one in three children under the age of five does not possess any record of their existence [14]. If every vaccinated child were to receive an identity, the 1.1 billion gap can be closed by a significant amount. By engaging trustworthy organizations such as primary healthcare centers, hospitals, educational institutes, etc. which interact with citizens on a regular basis as authorized identity-attesting entities, the resource burden on governments can be eased. Enabled by blockchain, one can envision an internet of identity where authorized institutions across the world can attest to the validity of identities, mathematical representations of which are recorded immutably on a public-permissioned blockchain.

Specifically, in the case of stateless and persecuted populations such as the Rohingyas, such a system holds immense value. Citizens are no longer dependent solely on one entity such as a national government to act as custodians of their identity. With the proof of identity stored immutably on the blockchain, citizens now have a persistent and portable identity which cannot be retrospectively erased by any single entity.

How would this function one might ask? By placing the user at the center of this system, one's identity then becomes a collection of identity claims attested by different entities. Imagine the case of a child whose parents generate an identity (a key pair) for the child on their smartphone and make claims about the child's identity such as name, date of birth, place of birth, etc. These claims may then be attested by the hospital, a proof of which is stored on the blockchain with the digitally signed attributes stored on the user's device. When the child enrolls to school, these attributes are shared with the school, which verifies that the identity attributes were attested by a hospital and then adds an additional attestation to the identity. The child's identity now has a stamp of approval from both a hospital and a school stored on a smartphone, the proof of authenticity which are verifiable on the blockchain. Thus, the identity grows in reliability with every additional use of the identity for authentication while the trail of use of this identity remains accessible only to the identity holder, thereby upholding privacy.

The challenges associated with this are rather apparent on first reading. If access to smartphones and reliable mobile internet is a prerequisite for possessing identities, what happens to those who lack access to them? If identity data is stored exclusively on the user's device, what happens to your identity if you lose your smartphone?

It is indeed true that the path to universal self-sovereign digital identity is a long and bumpy one. Solutions to the technical problems highlighted above are emerging, but still remain in its early stages. While we remain a long way from universal access to smartphones, one would be surprised to find that cer- tain sections among the most vulnerable populations who are excluded from identity programs do in fact possess smartphones. Stories abound of refugees who rely on smartphones to guide them through their treacherous journeys across continents and to find access to vital services [15]. In the absence of access to smartphones, smart cards, or other low-cost physical representations can serve as substitutes but have limited functionality compared to a smartphone. Innovators are finding methods of decentralized identity recovery where citizens can backup identities among multiple devices and also among trusted friends and family.

9.5 The State of Self-Sovereign Identity Today

While references to the concept which is today termed self-sovereign identity (SSI) had previously been explored among the cypherpunk community, the term self-sovereign identity may be attributed to Christopher Allen who published a blog post, "The Path to Self-Sovereign Identity" in 2016. The term self-sov- ereign is described by Christopher Allen as "an ideology to reclaim human dignity & authority in the digital world, and an emerging architecture of technology designed to enable that movement."

Since that start, SSI has come a long way. A set of technologies - the blockchain, decentralised identifiers (DIDs) and verifiable credentials – have been developed by the broader community to put SSI on a pathway to widespread adoption today.

Bolstered by these technological foundations, governments around the world are starting to take note and incorporate SSI into their digital identity programs. Europe has been a leader in driving this push. In 2018, the European Commision established the European Blockchain Services Infrastructure with the goal to provide a sandbox for testing a number of use cases with digital identity being a central enabler. In 2019, the European Self-Sovereign Identity Framework Lab (eSSIF-lab) was setup issuing grants to organisations and projects that could catalyse the development of SSI in the continent.

These initiatives have culminated with the publication of Europe's Digital Identity Regulation eIDAS 2.0 in May 2024 which strongly embraces self-sovereign principles & technologies with the goal to make such solutions accessible to all European residents by the end of 2026. Alongside Europe, a number of governments around the world including Canada, Switzerland, Bhutan have put citizen's privacy & control at the center of their digital identity programs, laying the foundations for a fairer digital identity future.

Method	DID prefix
Bitcoin Reference	did:btcr:
Blockstack	did:stack:
ERC725	did:erc725:
Ethereum uPort	did:uport:
IPFS	did:ipfs:
IPDB	did:ipdb:
Sovrin	did:sov:
Veres One	did:v1:

9.6 The Next Frontier in Self-Sovereign Digital Identity – Personal Data Management

With every click and keystroke on the web, to driving around in our hyperconnected cars we are giving away valuable personal data which constitute our digital identities. Large parts of this data is controlled and exploited by large corporations for their own profits. An evaluation of the most valuable companies in world, all data giants, serves as a reliable indicator of the value of our personal data.

While this has already been the reality for well over a decade, the advent of AI has only supercharged this phenomenon. Trained on user generated data, AI models are starting to excel at performing tasks previously thought to be achievable exclusively by humans. From creating art & music to making entire workplace functions obsolete, AI is already starting to transform society in profound ways. With AI advancing at a rapid pace, experts predict that Artificial General Intelligence – the hypothetical intelligence of a machine to understand or learn any intellectual task that a human being can – will be a reality within the next 2 decades [15].

While these models are trained on swathes of user-generated data, the monetary benefits accruing from these models are likely to be concentrated among those who own & control these models. By creating markets for this data, while putting the users that generate them in control of how such data is accessed and used, we could potentially mitigate the coming wave of job losses & inequality that AI could trigger.

Beyond the redistribution of wealth in the data economy that this could enable, personal data management platforms which place users in con- trol of their data answer to a more pressing need – the fundamental right to privacy. While the issue of ensuring absolute privacy in the digital realm extends far beyond the reaches of such solutions, putting ownership of data back in the hands of the user switches incentives for corporations whose busi- ness models rest solely upon intrusive data collection and monetization.

9.7 The State of Self-Sovereign Personal Data Management

The state of development of personal data management platforms are still rudimentary and the goals highlighted above are very lofty. In 2018, Europe once again led the way by bringing the General Data Protection Regulation (GDPR) into force. With far reaching powers for regulators to fine companies in breach of the law and mandates that companies were obliged by law to provide their personal data to citizens who request them, the GDPR set into motion a wave of privacy-focused innovation.

Starting 2018, personal data management platforms emerged, which allowed citizens to place data handover requests on smartphone apps and the ability to control & monetize this data if they so wished. However, these efforts failed to take off in a meaningful way given technological challenges impeding usability and a resulting lack of adoption among data consumers. A second wave of innovation focused on personal data management is now underway. Projects such as World Network aiming to create a privacy preserving global digital identity network which can eventually deliver Universal Basic Income to every human being in the world supported by Artificial Intelligence.

9.8 e-Voting

"If there were a nation of Gods, it would govern itself democratically. A government so perfect is not suited to men.", stated (Rousseau, 1762). A statement viewed as heretical at the time, seems much less so today. More than half the world's population lives in democratically governed countries. While the global acceptance of democracy has grown, little has changed in how voters exercise their democratic right of voting. Globally, only four nations use electronic voting on a nationwide scale [16].

In most parts of the world, citizens' role in governance is limited to making their choice known at the ballot every few years. In the interim, citizens remain powerless while their elected representatives make decisions on issues which have far-reaching consequences on their lives. Digital technologies have the potential to disrupt this status-quo and empower citizens with tools to participate actively in policy making.

From a technical perspective, voting is essentially a flow of information from the citizen to the government. With the internet and a smartphone, the cost of this information flow can be reduced significantly, making it economically feasible for governments to regularly elicit public opinion on consequential policy issues. As seen in Estonia, the introduction of the i-voting system which enables citizens to cast their votes through their smartphones saves the government 11,000 working days per election [17].

Despite the successes of Estonia, the rest of the world has been slow to embrace the possibilities brought to the table by i-voting. This may be attributed primarily to the fact that technologists have fallen short in convincing policymakers and the broader public that the technological resources at our disposal meet the security standards required to uphold the integrity of elections (Gritzalis, 2002).

Blockchain along with recent developments in cryptography hold the potential to change that. Blockchain based voting systems offer clear advantages over traditional systems which could help assuage concerns around security. Firstly, having every vote encrypted and recorded immutably on a publicly readable blockchain makes tampering with votes universally evident. This opens up the possibility of enabling voters to verify that their votes were cast as intended and counted as cast. Finally, homomorphic encryption makes the election result universally auditable without the need to decrypt individual votes.

9.9 The State of Blockchain-based eVoting Today

While the promise of blockchain-based eVoting is undeniable, and among the most valuable uses cases for blockchain at the moment, the road to mass adoption is a long one. Unlike the digital identity projects highlighted earlier in the chapter, whose measures of success are more long term, elections being specific events with a defined duration allow for a closer look at specific projects.

The quaint Swiss Canton of Zug which has positioned itself as a home for crypto and blockchain innovation in the recent past, once again took the pioneering step of experimenting with the use of blockchain for a non-binding vote on municipal policies. One of the issues put to the vote was whether citizens welcomed the use of blockchain-based eVoting in the Canton. With 100 residents of the Canton participating in the vote, 79% of the voters were in favour of the use of eVoting in the Canton [18]. The limited number of voters possibly makes the result biased owing to the fact that the vote was conducted digitally and those who are already comfortable using eVoting technologies, are more likely to vote favourable. Nonetheless, with the over- whelming majority of voters supporting the initiative, this may be viewed as a reflection of citizens seeing a need for more secure and convenient tools for citizen participation.

Elsewhere in the world, the US state of West Virginia allowed military personnel stationed abroad to vote through a mobile app in the November US midterm elections in 2018. This was following a successful trial of the same blockchain secured voting platform, allowing overseas military personnel to vote in the state's primary elections in April of the same year. While government authorities expressed a strong approval of the two trials, they also qualify that they have no near-term plans to expand the use of the blockchain secured voting plat- form beyond the current use [19][20] .

While these trials are hugely encouraging for the future of eVoting and consequently democracy, the technology is yet to be proven at scale. Blockchain-based eVoting is limited by many of the scalability challenges faced by public blockchains in general, solutions to which are emerging, yet remain in its early stages. Furthermore, most of the criticisms around blockchain-based eVoting state that while blockchain solves issues of auditability, some of the other challenges such as security vulnerabilities of smartphones and mobile internet networks still remain unsolved [21].

9.10 Exploring the Societal Implications of Digital Technologies

Our role as innovators however, does not stop at just building the technological foundations of democracy as we know it today but to use technology to challenge the boundaries of what it means to be a citizen. At a time when democracy is facing a multi-pronged threat from rising populism, challenges to free speech, growing polarisation in public opinion aggravated by algorithms and fake news, one needs to step back and take a holistic view on society and the zeitgeist to inform the process of innovation.

Building the technological foundations of tomorrow's society cannot exclusively be the job of technologists. Greater use of technology or digitization is not an end in itself, but merely a tool which can drive positive social outcomes if designed right. Society needs forums that bring together a diverse pool of experts to broaden the debates and conduct research which steers innovation toward favourable social outcomes. As we have seen in the past, innovation is inevitably followed by unin- tended consequences. Social media platforms, which were seen as torch- bearers of free speech and democracy, empowering the marginalized with voices now threaten the very foundations of democracy. This has led many prominent voices to oppose innovation and seek refuge in the security of the known. But as history has shown repeatedly, innovation and change cannot be stopped, rather, they are the only constants which underpin society. The answer lies not in thwarting innovation but in engaging diverse voices to integrate broader social and environmental cognizance into innovation.

New technologies pose more questions than answers. The answers to these questions might lie in unexpected places, perhaps with you. Building a Humanized Internet cannot be the job of a select few but that of everyone impacted by it. This chapter is thus not intended to serve as an account of our work but rather to spark the interest of you, our readers to get involved in the process of innovation. The spirit of the Humanized Internet is just this – leveraging the wisdom of the collective to build technology which redistrib- utes power equitably to all, including the most vulnerable.

References

[1] https://www.repository.cam.ac.uk/bitstream/handle/1810/247570/Bell%202014%20Political%20Theory.pdf?sequence=1
[2] https://www.economist.com/schools-brief/2018/08/11/de-tocqueville-and-the-french-exception
[3] https://ourworldindata.org/democracy
[4] https://data.worldbank.org/topic/poverty
[5] http://blogs.worldbank.org/developmenttalk/node/1140
[6] https://d1tn3vj7xz9fdh.cloudfront.net/s3fs-public/file_attachments/bp210-economy-one-percent-tax-havens-180116-en_0.pdf
[7] https://www.economist.com/bagehots-notebook/2018/06/12/some-thoughts-on-the-crisis-of-liberalism-and-how-to-fix-it
[8] https://e-estonia.com/solutions/e-governance/
[9] https://uidai.gov.in/aadhaar_dashboard/
[10] https://pmjdy.gov.in/account

[11] https://timesofindia.indiatimes.com/india/supreme-court-reserves-verdict-on-aadhaar-validity/articleshow/64116972.cms
[12] https://www.itu.int/en/ITU-T/focusgroups/dfs/Documents/09_2016/Review%20of%20National%20Identity%20Programs.pdf
[13] https://www.technologyreview.com/s/604294/solving-a-global-digital-identity-crisis/?set=604287
[14] https://www.economist.com/international/2017/02/11/phones-are-now-indispensable-for-refugees
[15] https://cloud.google.com/discover/what-is-artificial-general-intelligence
[16] https://economictimes.indiatimes.com/news/politics-and-nation/this-is-how-evms-win-voter-confidence/how-many-countries-use-electronic-voting/slideshow/64159925.cms
[17] https://e-estonia.com/solutions/e-governance/i-voting/
[18] https://cryptovalley.swiss/citizens-open-more-blockchain-evoting/
[19] https://cointelegraph.com/news/west-virginia-secretary-of-state-reports-successful-blockchain-voting-in-2018-midterm-elections
[20] https://cointelegraph.com/news/blockchain-and-elections-the-japanese-swiss-and-american-experience
[21] https://www.technologyreview.com/s/611850/why-security-experts-hate-that-blockchain-voting-will-be-used-in-the-midterm-elections/

Conclusion

In an era where digital technologies permeate every facet of our lives, understanding the interplay between our digital identities, personal dignity, and democratic values has never been more critical. "The Humanized Internet "has endeavored to illuminate the complex and often precarious relationship we maintain with the online world, where our digital selves can empower and expose us. As we navigate this shifting landscape, this book beckons us to reflect on the fundamental principles that should guide our digital interactions and the governance of the internet. Throughout our exploration, we have witnessed how the internet, originally envisaged as a tool for connection and equality, has, over time been subverted into a medium where personal data is commodified, privacy is often an afterthought, and misinformation can easily dominate the discourse.

This transformation poses significant threats to individual dignity; the notion that every person is worthy of respect and ethical consideration. In confronting these challenges, it becomes imperative that we advocate for a digital environment that honors individual dignity and promotes equitable access to information and representation. Moreover, the book underscores the necessity for a reevaluation of our digital identities. As users of the internet, we are not merely consumers of content; we are active participants in shaping the narratives that define our virtual communities and, by extension, our societies. This agency carries with it the responsibility to engage critically with the platforms we inhabit and to demand greater accountability from those who govern these digital realms. The call for transparency and ethical practices within tech companies is not merely a movement for better conduct; it is a foundational requirement for a functioning democracy in the digital age.

Democracy itself hinges on the informed and active citizenry. Therefore, cultivating a humanized internet is not just about safeguarding dignity but also ensuring that democratic ideals are upheld. When individuals can express

DOI: 10.1201/9781003339724-15

themselves freely, access diverse perspectives, and participate in meaningful dialogue, the foundations of democracy are strengthened. The interplay between the digital sphere and democratic engagement is a bidirectional one; as we nurture a more inclusive and just online environment, we also reinforce the principles that underpin our democratic institutions.

In conclusion, "The Humanized Internet" serves as a clarion call not just for policymakers, technologists, and advocates, but for all individuals who navigate the digital landscape. It invites us to envision an internet that serves as a facilitator of dignity, and a guardian of democratic ideals. As we move forward, let us commit to harnessing the potential of our interconnected world in ways that respect individual dignity and bolster the democratic process. Only then can we truly realize the promise if a Humanized Internet, one where technology empowers rather than diminishes where every voice us valued, and where every individual can thrive in both digital and democratic realms.

Additional Research

There are opportunities for additional research such as:

1. User centric design principles establish frameworks for designing internet interfaces and services that prioritize user empathy and accessibility. Research should explore how these principles can be systematically implemented across various platforms to enhance user satisfaction and engagement.
2. Impact assessment investigates the social, psychological, economic impacts of the Humanized Internet. This includes studying how human-centered technologies influence mental health, social interactions, and community-building. Assessing these impacts can provide insights into the benefits and potential drawbacks of such technologies.
3. Ethics and governance examine the ethical dimensions and governance structures surrounding the Humanized Internet. This research should focus on privacy concerns, data security and the implications of algorithmic bias. Establishing robust ethical guidelines will help foster community trust and safeguard user rights.
4. Cultural sensitivity and diversity explore how the Humanized Internet can cater to diverse cultural contexts and populations. Research should investigate cross-cultural design methods and strategies for incorporating diverse perspectives and needs into the internet technologies.
5. Interdisciplinary collaboration encourages partnerships between technologists, social scientists, ethicists, and policymakers. Collaborative research can produce more holistic solutions that consider technical, social, and ethical factors impacting the Humanized Internet.

Additional research into the Humanized Internet is critical for addressing the pressing challenges of our digital world. The outcome of such research has the potential to significantly enhance the quality of digital interactions that ensure technology serves humanity effectively and compassionately.

Some *recommended actions* include the following:

1. Establish dedicated funding opportunities for Humanized Internet research initiatives.
2. Facilitate workshops and seminars to discuss ongoing research findings and their implications.
3. Create a consortium of stakeholders, including technologists, researchers and policymakers to drive advancements in this field.

By prioritizing these research areas, we can contribute to a future where technology is more harmoniously aligned with human needs and aspirations.

Index

About the Authors

Akram is a member of The Humanized Internet. Akram's work is rooted in the intersection of Technology, social work, and human rights. Since early 2016, Akram has been actively engaged in the early stage of several initiatives in the 'Tech for refugees' scene in Germany. His motivation to apply technology solutions toward the betterment of our society and the fact that he has personally experienced defects of the current identity political system made him join The Humanized Internet. Akram proposes a more fundamental approach to help those in need, as he believes that "empowering" only starts by guaranteeing basic human rights, and that identity comes in the center of any substantial solution.

With over 25 years of experience as a global technology leader, **Monique Morrow** has established a proven track record of success in driving business growth and innovation through emerging technologies.

She holds several industry advisory positions and most recently Venture Partner Sparklabs Accelerator for Cybersecurity and Blockchain, Wash D.C. Monique is also an Independent Director on the Hedera Hashgraph Board of Directors where she focuses on articulating Hedera enterprise value for the telecommunications industry.

In Monique's previous role as Senior Distinguished Architect for emerging technologies at Syniverse Technologies, she provided thought leadership and developed the strategic direction and vision for Syniverse's emerging technologies across the company, partners, and industry forum.

Monique's expertise is in cybersecurity, privacy, mobile payments, ethics in extended reality, and distributed ledger technologies, for example, blockchain.

She has worked across the industry including AMD, Ascom Hasler, Swisscom and Cisco, and has been active in the start-up community. Additionally, Monique is president and co-founder of the Humanized Internet a Swiss-based non-profit with a focus on digital identity, and ethics in technology. Monique is an IEEE Senior Member.

Monique's industry leadership includes the following examples:

Monique chairs the IEEE P7030 Global Extended Reality Working Group, Recommended practice for the assessment of Extended Reality Technologies. She also Chairs IEEE P7016 "Standard for Ethically Aligned Design and Operation of Metaverse Systems." Monique was Co-Chair of the GSMA-Distributed Ledger Technology [DLT] group and served as Syniverse's representative in the World Economic Forum [WEF] Data Policy Council and Digital Justice. Monique had contributed to the WEF-Wharton School Paper on the "Decentralized Autonomous Organization Toolkit." She was active in WEF's Digital Identity initiative. Monique also contributed to the GSMA, Vodafone, IBM Chaired white paper entitled, "Post Quantum Telco Network Impact Assessment."

Monique has been recognized in the industry for her tireless focus on social good. She holds over 17 patents and has co-authored several books. Amongst Monique's accomplishments include the following: Monique was selected as one of the top Digital Shapers 2018 in Switzerland. Forbes Magazine listed Monique Morrow as one of the top 50 women globally in technology. OneWorldIdentity recognized Monique as one of the top 100 influencers in identity for 2019. In May 2019, Monique was recognized by Cybersecurity Ventures as one of the top 100 women in Cybersecurity. In March 2020, Monique was one of several innovators featured in Red Bull Innovator Magazine. In April 2020, Monique was selected as one of the 5 leading figures in Business by the Europa Forum Lucerne (Switzerland) specific to its theme, "Safety and Security in Times of Uncertainty." Monique is also the 2020 winner of the WomenTech Network Global Technology Leadership award. Monique has been recognized in 2021 as one of the top 100 Women in Cybersecurity in Europe. Monique was named by Chief in Tech as one of the top 100 Women in Tech Leaders to watch in 2022. August 2024, Monique was recognized as one of the top 150 women fighting Cybercrime.

Connect with Monique Morrow on LinkedIn

For Product Safety Concerns and Information please contact our EU
representative GPSR@taylorandfrancis.com
Taylor & Francis Verlag GmbH, Kaufingerstraße 24, 80331 München, Germany

www.ingramcontent.com/pod-product-compliance
Ingram Content Group UK Ltd.
Pitfield, Milton Keynes, MK11 3LW, UK
UKHW022211190926
475702UK00003B/305
* 9 7 8 8 7 7 0 2 2 0 3 2 3 *